ORIGINS OF SCIENTIFIC LEARNING

ORIGINS OF SCIENTIFIC LEARNING

Essays on Culture and Knowledge in Early Modern Europe

Edited by
Sara L. French
and
Kay Etheridge

With a Foreword by
Jonathan Seitz

The Edwin Mellen Press
Lewiston•Queenston•Lampeter

Library of Congress Cataloging-in-Publication Data

Origins of scientific learning : essays on culture and knowledge in early Modern Europe / edited by Sara L. French and Kay Etheridge ; with a foreword by Jonathan Seitz.
p. cm.
Includes bibliographical references.
ISBN-13: 978-0-7734-5369-2
ISBN-10: 0-7734-5369-5
1. Science--Europe--History. 2. Science--Europe--Philosophy. 3. Philosophy, European--History. I. French, Sara L. (Sara Lillian), 1968- II. Etheridge, Kay.
Q127.E8O75 2007
509.409'03--dc22

2007031293

hors série.

A CIP catalog record for this book is available from the British Library.

Front cover: Frontispiece from Imperato, Ferrante. *Dell'historia Naturale.* Napoli: C. Vitale, 1599. Image courtesy of the Department of Library Services, American Museum of Natural History.

The Edwin Mellen Press
Box 450
Lewiston, New York
USA 14092-0450

The Edwin Mellen Press
Box 67
Queenston, Ontario
CANADA L0S 1L0

The Edwin Mellen Press, Ltd.
Lampeter, Ceredigion, Wales
UNITED KINGDOM SA48 8LT

Printed in the United States of America

To the memory of Esther Griswold French, SLF

To my parents, Sara and Ray Etheridge, KE

CONTENTS

ILLUSTRATIONS

FOREWORD

Scholars have identified in Early Modern Europe dramatic transformations in a wide array of human endeavors: political, artistic, literary, scientific and technological, among others. However, contemporary scholarship has only recently begun to cross disciplinary boundaries to tie changes in literature, for instance, to those in technology. The papers in this volume contribute to this interdisciplinary effort, drawing connections among Early Modern metamorphoses often studied in isolation.

These papers arose from a conference sponsored by Binghamton University's Center for Medieval and Renaissance Studies in 2004. The conference, organized by Dana Stewart around the theme of "Science, Literature, and the Arts in the Medieval and Early Modern World," showcased the diversity of work being done on medieval and Early Modern Europe. All but one of the essays presented here are revised and extended versions of papers delivered at the Binghamton conference.

The "science" discussed in this volume is science in its broadest sense—including the study of the natural world in all its aspects (including the human body) as well as relationships between humans and their natural and built environments. The scholars included here reflect the present state of Early Modern studies and the great variety of interests and approaches Early Modernists use, from the approaches and tools of literary criticism to those of philosophy to those of history, among others.

Kay Etheridge, for example, brings a biologist's eye to the examination of images of reptiles and amphibians in Early Modern natural-historical texts. She highlights the trend toward naturalism in many of these works, but also points out that these images could provide information to Early Modern naturalists that was superior to the information that could be gained from preserved specimens. Building on William Ashworth's work, Etheridge traces the illustrators' motives as they shifted from an animal's symbolic connections to illustrating its place in the emerging catalog of nature. But the development of these images did not end there. Indeed, as Etheridge shows, artist-naturalists like Maria Sibylla Merian began to transcend mere cataloging to *contextualize* specimens, depicting creatures surrounded by related flora and fauna.

Natural history is also the arena for Michael Cunningham's examination of the work of Antonio Vallisneri, a naturalist fascinated by fossils' potential connections to the biblical story of Noah's flood. Vallisneri's discussion of the natural evidence of the Deluge involved him perhaps even more deeply than Galileo Galilei in the thorny relationship between scriptural and natural knowledge. In Cunningham's account, Vallisneri is revealed as a natural philosopher emerging from the epistemic mold of Galileo and treading carefully between natural and revealed knowledge. Vallisneri was perfectly willing to reinterpret scripture in light of his natural historical conclusions, but he was far from the libertine's rejection of revelation in favor of reason. Indeed, Vallisneri made use of literal interpretations of scripture when they harmonized with his natural knowledge—as in the case of the centuries-long life-spans of biblical figures which he felt supported his contention that the Earth was far older than most of his contemporaries believed.

Daniel Selcer, in turn, grounds his study in the history of philosophy, exploring the methodological tensions between analytical and historicist approaches in that field. As a case study, Selcer uses the "Système Nouveau" of Gottfried Leibniz, and he turns the ideas of materiality expressed in Leibniz's philosophical text onto the text itself, which is in its own way an "aggregate" just as are objects in the corpuscular mechanical philosophy to which Leibniz adhered.

Consonant with the call of the original conference, a number of the essays presented here go beyond the traditional bounds of the history of science to pay special attention to the links between science and literature in Early Modern Europe. The contributions of Elizabeth Oldman and Lauren Klein both examine how technology was involved with the literary devices of Early Modern authors.

After the Thirty Years' War (1618-1648) demonstrated the increased lethality and destructive power of Early Modern warfare—due in part to the ubiquity of firearms—existing ideas about justice and war came under increasing pressure. Oldman traces how these concepts developed in the legal writings of Hugo Grotius and in John Milton's *Paradise Lost*. Grotius, in such writings as his *De jure belli ac paci*, argued that war undertaken for proper motives and waged in accordance with natural law could indeed be just. Milton offers a parallel vision, Oldman argues, in his description of the War in Heaven, wherein Satan's unrestrained, gunpowder-fired attack is nevertheless defeated by the more narrowly targeted tactics and "natural" weapons of God, the Son, and the faithful angels. Hence, ambivalence about this emergent technology is revealed not only in legal considerations of warfare, but in poetical ones as well.

In Klein's case, the key text is Francis Bacon's *Great Instauration*. Klein suggests that Bacon's repeated use of images and metaphors of mining in his epistemological arguments is puzzling, given the negative connotation of mining activities in his day. What did Bacon find attractive in the metaphor of the mine? To solve this puzzle, Klein considers the metaphor's value as a literary device, while also drawing on Evelyn Fox Keller's analysis of Bacon's construction of a distinctly "masculine" methodology. Given that mining was, at times, described as a violation of the feminine Earth or Nature, Klein shows such a metaphor to be quite in character for Bacon. Moreover, the mining metaphor allows Bacon to describe the advance from raw (observational) ore to refined (theoretical) products shaped by human intervention into forms not found in nature — supporting his desire to combine empirical and experimental investigation in order to yield inductive generalizations.

Klein's work highlights for us the frequent implication of gender in discussions of science in the Early Modern period. Indeed, the question of what makes male and female, and the permeability of the two categories, is taken up by several authors in this volume. Like Oldman and Klein, Roberta Martin places literary trends in their scientific context, here connecting poetry to the Early Modern "culture of dissection" that quite literally opened up the body to increased observation. Drawing on recent work in the social history of medicine, Martin contrasts the poetry of John Donne with that of Aphra Behn to reveal the power that Early Modern Europeans believed dismemberment (metaphorical or otherwise) gave the observer over the observed. Donne figuratively flays his lover, which allows him to lay claim to her. In contrast, Behn's shepherd Lysander fails to dismember *his* would-be lover and is thus rendered impotent. Behn proceeds to co-opt traditional humoral medical theory—a medical theory that privileges the male as Thomas Laqueur has argued—to re-cast the shepherd Lysander as a female and blur the boundaries between the sexes.

Female authorship and the medical division of the sexes is also at the core of Beatriz Cruz-Sotomayor's examination of a sixteenth-century Spanish treatise in which the subversive feminine and dominant masculine combine to produce a text that is at once deferent and rebellious. Cruz-Sotomayor notes that even as the text suggests a novel physiology and draws power from its (at least putative) female authorship, it also reiterates aspects of medical theory it challenges—including elements that express the inferiority of the feminine, such as the supposed destructive powers of menstrual blood.

In her contribution, Sara French explores an episode in which we find challenged not the corporeal basis of gender divisions, but rather their instantiation in built environments. As French demonstrates, just as Queen Elizabeth's lengthy reign in England challenged dominant assumptions about women's capabilities and (figurative) place in society, so the rearrangement of the English country house by Bess of Hardwick challenged the (literal) place of women in the manor. Bess devalued the traditional great hall, a masculine space for the display of lordly power, and made it a mere extension of serving spaces. In its place, Bess's state rooms became the location for the exercise of power in Hardwick Hall. Thus did both Bess and Elizabeth challenge the pre-modern conception the female as an imperfect, inferior version of the male.

The remarkable coverage of the essays in this volume is a testament to the broadening of Early Modern studies, and the productivity of interdisciplinary analyses of the period. The days of the history of science and medicine narrowly conceived as the internal history of ideas separated from the broader history of the period are, thankfully, past, and have been replaced by much richer conversations among historians, philosophers, literary scholars, art historians and other scholars. The increasing use of new histories of science and medicine by the broader Early Modernist community allows a fuller contextualization of art, literature and law, to take just a few of the topics presented in these pages. At the same time, new tools are brought to the analysis of scientific texts and ideas, recovering the role of women in medical debates, exposing the literary techniques of natural philosophers, and tracing the spread of "scientific" ideas and attitudes into the broader culture.

The authors here reveal the vibrancy of contemporary Early Modern studies, and the fruitfulness that methodological and disciplinary cross-fertilization yields. The essays serve as a reminder to expand one's scholarly horizons beyond the sources, methods and topics of one's own corner of history to embrace what can be learned about gender from architecture or technology from poetry.

Jonathan Seitz
Drexel University

Jonathan Seitz received his Ph.D. from the University of Wisconsin—Madison and is presently on the faculty at Drexel University. His research focuses on conceptions of nature and the supernatural in Early Modern Europe, and he is currently completing a study of how physicians, clerics and ordinary people identified and understood witchcraft and other unnatural illnesses in sixteenth- and seventeenth-century Venice.

ACKNOWLEDGMENTS

Sara L. French would like to thank the Faculty of Landscape Architecture at the State University of New York College of Environmental Science & Forestry for their support during the editing of this volume. Individual colleagues have also been instrumental in providing their professional expertise, including Elen Deming, Thaïsa Way, and Katherine Auwaerter. Thanks are also due to David Patterson for his assistance in preparing the final illustrations. A final thank you is due to Bob Cundall for his support and assistance in every crisis.

Kay Etheridge would like to thank the Gettysburg College Provost's office and the Biology Department for financial support of travel and the costs of reproducing the figures in "Loathsome Beasts." She is also very grateful to many very helpful and knowledgeable librarians, including Tom Baione, Leslie Overstreet, Daria Wingreen, Kim Sloan.

The Edges of Extension and the Limits of the Text: Leibniz, Materiality, and History

Daniel Selcer

This essay is an attempt to map the contours of a disciplinary disconnection and methodological challenge facing scholars of the Renaissance, early modernity, and the Baroque. Its focus is the question of how we are to negotiate the apparent divide between philosophical conceptuality and historicism. More specifically, I have in mind the tension between historicist (particularly new historicist) and microhistorical approaches on the one hand and more directly theoretical or philosophical interpretive methodologies on the other. In the following, I do not claim to resolve the tensions between these approaches by finding a middle way between materialist historicism (not to be confused with historical materialism) and philosophical-conceptual interpretation. I do, however, attempt to bring elements of the arsenals of each to bear on the other by examining an early modern philosophical work exemplary for its amenity to both, the philosopher Gottfried Wilhelm Leibniz's text, "Système nouveau de la nature et de la communication des substances, aussi bien que de l'union qu'il y a entre l'âme et le corps." I am not proposing that this text holds any secret resources for resolving the history-theory problem and that everyone should therefore pay attention to Leibniz. I am using it as an example because in my own teaching and writing it is one of the works that for me has posed most starkly the interpretive and methodological dilemma that has materialist historicism as one of its horns and conceptual interpretation as the other.

As a first step, I will outline what I take to be the nature of this impasse. I will then briefly discuss Leibniz's "Système nouveau" as a discrete textual and material object with a particular history of circulation and interpretation. Finally, I will turn to one element of the philosophical argument of the text — its theory of materiality involving the stance that bodies lack definite edges and limits — in an effort to suggest that this conceptual move might have productive repercussions when applied to the text itself. Ultimately, I want to try to describe the problem of the material limits of this text's extent (the contours of what counts, historically, *as* the "Système nouveau") while also engaging its critique of extension as adequate to materiality (the contours of Leibniz's attack on Cartesian corporeality *within* the "Système nouveau"). If the limits of this early modern philosophical text are related to its materiality and history, then it might be helpful to turn to the theory of materiality it contains in order to understand those limits. Conversely, an analysis of the amorphous constitution of its own limits as a historical text ought to be able to inform our interpretation of the theory of corporeality it articulates.

At many recent interdisciplinary meetings dealing with the Renaissance and early modernity, complaints about 'mere historicism' have been on the rise. Some scholars seem to be expressing a yearning for a return to the theory (pure or otherwise) that has fallen to the wayside due to the current focus in most humanistic disciplines on historical particularity and material culture. In the August 2004 special issue of *Critical Inquiry* focused on the "Arts of Transmission," one finds many of the most prominent practitioners of materialist historiography uncharacteristically struggling with the problem of connecting their historical work with broader philosophical or 'purely theoretical' questions.[1] I do not mean to claim that these scholars have not drawn such connections before; indeed, they are among the most deft and rigorous scholars concerned with such connections writing today. Rather, the essays in this issue display a tension that one rarely finds in many of their authors' work, as if the call for a broad and theoretical consideration of the way that technologies for the transmission of knowledge (whether theoretical or practical, intellectual or embodied) impact and affect the contours of that knowledge had made explicit a

[1] "Arts of Transmission" special issue, ed. James Chandler, Arnold Davidson, and Adrian Johns, *Critical Inquiry* 31:1 (August 2004).

problem the authors generally dealt with implicitly. The problem is that if one is truly committed to a materialist view of literary, intellectual, philosophical, and scientific history, then the theoretical *implications* of one's own historical work can be regarded as having little more real status than the sloughed-off *eidola* identified by one of the original materialists, Epicurus.[2] Epicurus saw these *eidola* as the fine remnants of the shapes of existent things — constituted by bodies and void alone — left behind like a sticky residue as the constituent matter of objects moved or changed over time. Just as Epicurus held that human perception depended on the organs of sense accidentally encountering these fragile and fleeting *eidola* as we moved through the world, so too do we encounter the implicit philosophical and theoretical propositions that animate the passions of materialist historiography. Likewise, just as positively encountering the *phantasia* of *eidola* meant, for Epicurus, that we never perceived bodies themselves except through corporeal collision, so too does the *implicit* nature of such theorization defend it from being forced to interrogate its own grounds.

It is probably not legitimate to identify microhistoricism itself as the real terrain of these tensions and anecdotal complaints, certainly not in the form practiced either by those scholars mentioned above or — in its 'pure' form — by writers like Carlo Ginzburg and Anthony Grafton. As most of those familiar with this discourse will know, Ginzburg's *The Cheese and the Worms*, originally published in 1976, is often taken to be one of the initiating texts of the microhistorical discourse.[3] Ginzburg's methods, of course, stand in a lineage of direct descent from and resistance to the first- and second-wave *Annales* School historians. Where the practitioners of that largely French movement sought to

---

[2] See Epicurus to Herodotus, §46-52 in *The Epicurus Reader*, trans. and ed. Brad Inwood and L.P. Gerson (Indianapolis: Hackett Publishing, 1994), 5-19.

[3] Carlo Ginzburg, *The Cheese and the Worms: The Cosmos of a Sixteenth-Century Miller*, trans. John and Anne Tedeschi (New York: Penguin, 1980). Also see Ginzburg's own account of the origins of the terminology associated with microhistory in Carlo Ginzburg, "Micro History: Two or Three Things That I Know About It," trans. John Tedeschi and Anne C. Tedeschi, *Critical Inquiry* 20:1 (Autumn 1993): 10-35.

organize the practice of history through massive synchronic, serial projects investigating the records of everyday life rather than a narrative of 'world historical' events or individuals, Ginzburg transforms their archival rigor and emphases on the role of everyday life into micro-investigations of 'everyday' individuals and small communities. The diaries and business records of a single human being stand as the foundation for a reconstructive account of historical life. What is sought, in other words, is a historical account of the everyday fabric of lives, spaces, and practices that are no longer directly accessible to us. Grafton, by contrast, attempts to shift microhistorical methodology into the sphere of intellectual history. His work seeks new modes of access to the philosophical, theological, and religious positions of the Renaissance and early modernity by applying the same kind of focused attention in two ways. On the one hand, some of his work concentrates on figures that normally slip below the radar of philosophers and intellectual historians in order to give an account of what one might call the metaphysics or theory of everyday intellectual life, as opposed to the way it acquires the weight of tradition and interpretation in canonical texts. On the other hand, Grafton also engages intellectual giants of the period, but does so in order to illuminate their philosophical and scientific *practices* of research, writing, and dialogue, usually by concentrating on the way we can understand them to be readers (and concomitantly, what *this* can tell us about the history of practices of reading).[4]

I think that what is beginning to make some scholars uneasy about materialist historiography could be more legitimately described as the particular forms of microhistory that have been associated with the dominant style of that relatively amorphous method of literary and cultural history known as new historicism (a label rejected even by its purported founder, Stephen Greenblatt, who prefers instead to use "cultural poetics," a term evoking an entirely different

---

[4] Anthony Grafton, *Cardano's Cosmos: The Worlds and Works of a Renaissance Astrologer* (Cambridge, MA.: Harvard University Press, 1999). "Johannes Kepler: The New Astronomer Reads Ancient Texts" in *Commerce With the Classics: Ancient Books and Renaissance Readers* (Ann Arbor: University of Michigan Press, 1997).

set of theoretical resonances).[5] What is particularly significant about new historicist approaches is that in tandem with microhistorical methodologies in the social sciences, their rather vague set of premises now constitute the *de facto* and dominant methodological ground and theoretical standpoint in the humanities (with the exception of the rare congruence of departments of history, art history, and philosophy).[6] Even while a rash of publications over the last decade have attempted to debate the terms of the new historicist tenets (usually concluding with the alternately rigorous and honest or coy and fatuous claim that it has none), I think a number of elements of its theoretical orientation can be identified.[7] I would like to accord these tenets the status of 'suspicions' rather than premises, as most of those who identify themselves with the approach subscribe to some and not others. These suspicions include: the rejection of the classical privilege accorded to philosophy and intellectual history as totalizing theoretical and historical narratives; a concordant insistence that formal texts can only be understood in terms of their immersion and circulation in a material and ineluctably political culture; an emphasis on the legibility or textualization of non-textual historical and cultural artifacts and their consequent amenability to interpretation; a tendency to accord metaphysical privilege to the singularity and

---

[5] If new historicism has a founding document, it is Stephen Greenblatt's *Renaissance Self-Fashioning: From Moore to Shakespeare* (Chicago: University of Chicago Press, 1980).

[6] What philosophers and art historians share here is an insistence on the reality of their disciplinary objects. Philosophers generally feel they cannot do without real philosophical concepts; art historians need real art objects. Most historians proper, meanwhile, resist the textualization demanded by New Historical approaches. It is of course true that some important theoretical and methodological orientations in all three disciplines reject these premises. It is equally true, I propose, that typical scholars in these fields today do not.

[7] A sampling of the broad field encompassing such attempts (all listed here fall into the 'rigorous and honest' category) would include: Claire Colebrook, *New Literary Histories: New Historicism and Contemporary Criticism* (Manchester: Manchester University Press, 1997); Catherine Gallagher and Stephen Greenblatt, *Practicing New Historicism* (Chicago: University of Chicago Press, 2000); Ivo Kamps, "Materialist Shakespeare: An Introduction," in *Materialist Shakespeare: A History*, ed. Ivo Kamps (London: Verso, 1995), 1-19; and the essays collected in *The New Historicism*, ed. H. Aram Veeser (London: Routledge, 1989) as well as *The New Historicism Reader*, ed. H. Aram Veeser (London: Routledge, 1994). See also James J. Paxson, "The Green[blatt]ing of America: Reflections on the Institutional Genealogy of Greenblatt's New Historicism," *The Minnesota Review* n.s. 41-42 (1993): 221-35.

particularity over multiplicity and universality; and finally, an implicit leap from this singularity and particularity to more general claims about synchronic epistemic structures of ideology and power that govern the production of meaning and practice in any given micro-domain of investigation.

Many will surely protest that the elements I have just described distort the true contours of new historicism in its contact with microhistory, or at least that neither is as monolithic or systematic an approach as to be captured even by this loose description. I am willing to admit that both those things are the case. I think these readers would probably *also* agree, however, that this list of quasi-premises does go some distance toward describing a more or less dominant trend in the research and paper-writing practices of many who identify themselves partisans of these methods. At the very least, I think it describes the dominant premises entrenched in contemporary graduate education in most humanistic disciplines.

Within the discipline of philosophy, methodological evaluations of the proper level of attention toward the historicity of discourse and its textual instantiations are quite different from those of the other humanistic disciplines. In general, historians of philosophy are reluctant to embrace a view that subordinates direct conceptual engagement with ideas to their historical contextualization. Partly, this can be explained by the way philosophers in the Anglo-American context understood the nature of the discipline itself, at least for much of the twentieth century. Whether grasped in terms of the primacy of linguistic analysis or a pragmatics of interpretive practices, philosophers generally understood their work to have some connection either to the establishment of the grounds for claims that are atemporally and ahistorically logically true or coherent or, at the very least, to the debunking of claims that are neither of these. With the rise to institutional primacy of logical positivism and its broader array of analytic successors, for much of this period the most central 'historical' question asked by philosophers was whether the history of philosophy actually had a place within the discipline or whether it merely constituted the annals of philosophical errors. Throughout the 1980s, institutional debates over the possibility of a "plurality" of philosophical approaches wracked the discipline, often seemingly taking the form of a struggle for the meager power and resources of the American Philosophical Association among members of competing institutions of higher learning. This "pluralism debate" is usually understood in terms of the relationship between the so-called Analytic and Continental approaches, and the conflict between these schools was a dominant aspect of that struggle. The question of whether the discipline was open to a plurality of approaches, however, also reflected a shift in the practices of scholars working in the history of philosophy, particularly (at that specific point) those focused on early modernity and the Enlightenment.

In recent decades a vigorous debate has been underway among historians of philosophy as to the proper approach to their practice. Attention to the centrality of scientific concerns, struggles over religious dogma and its relationship to philosophical writing, and the general historical contextualization of philosophical texts have wrought a methodological transformation among many historians of early modern philosophy. These scholars are interested in engaging the broader intellectual milieu for the production of philosophical notions, and reject the narrative in which the history of philosophy is valued only insofar as it can contribute to filling in the background of 'contemporary philosophical problems' or insofar as it makes plain the series of 'wrong turns' or 'mistaken paths' that earlier philosophers have followed. Most contemporary historians of philosophy no longer defend the notion that only a select group of philosophical heroes is worth interpretive attention, eschew the claim that the rational reconstruction of historical arguments is an end sufficient unto itself, and strive to avoid imposing an anachronistic vocabulary on the texts they interpret.[8]

---

[8] Important texts in this regard include: Daniel Garber, "Does History Have a Future? Some Reflections on Bennett and Doing Philosophy Historically" in *Descartes Embodied: Reading Cartesian Philosophy Through Cartesian Science* (Cambridge: Cambridge University Press, 2001); Jill Kray, Introduction to *Classical Traditions in Renaissance Philosophy* (Aldershot: Ashgate, 2002); Christia Mercer and Eileen O'Neill, Introduction to *Early Modern Philosophy: Mind, Matter, and Metaphysics* (Oxford: Oxford University Press, 2005); Margaret J. Osler, "The *History* of Philosophy and the History of *Philosophy*: A Plea for Textual History in Context," *Journal of the History of Philosophy* 40:4 (Oct. 2002): 529-33; Richard Popkin, "Philosophy and the History of Philosophy" in *The Third Force in Seventeenth-Century Thought* (Leiden: Brill, 1992); and Richard Rorty, "The Historiography of Philosophy: Four Genres," in

As this shift took place, the debate largely revolved around questions of the legitimation of a contextualist view of the history of philosophy over and against so-called "analytic history of philosophy" and was centered largely on interpretive questions regarding early modern philosophical texts. The analytic view, most distinctly instantiated in the work of Jonathan Bennett, holds that the contextual meaning and intention of historical authors of philosophical texts is largely if not entirely irrelevant to their contemporary interpretation.[9] Analytic historians of philosophy viewed the textual history of their discipline as a mere adjunct to the formulation, definition, and solution of problems with which contemporary philosophers were concerned. Whether or not a particular philosopher ever held or intended to hold a particular view was taken to be largely beside the point. Insofar as that question entered into the consideration of an analytic historian, the interpretive endeavor was to determine whether the historical philosophers were, in fact, correct, and whether the arguments they presented in support of their views were internally and externally consistent. The question, as Kant once put it, was not what a philosopher actually said, but what they meant to say.[10] It followed that analytic work in the history of philosophy was to be sharply distinguished from intellectual history that, while interesting in its own right, contributed little to the ongoing development of philosophical discourse itself. Where intellectual historians primarily cared about the views that philosophers of the past actually held and why they held them, the task of analytic history of philosophy was to judge whether or not these philosophers were correct

Philosophy in History: Essays on the Historiography of Philosophy, ed. Rorty, J. B. Schneewind, and Quentin Skinner (Cambridge: Cambridge University Press, 1984).

[9] See especially Jonathan Bennett, *A Study of Spinoza's 'Ethics'* (Indianapolis: Hackett Publishing, 1984).

[10] Immanuel Kant, "Über eine Entdeckung nach der alle neue Kritik der reinen Vernunft durch eine ältere Entbehrlich gemacht werden soll" in *Werkausgabe*, ed. Wilhelm Weischedel, 12 vols. (Frankfurt am Main: Suhrkamp, 1996), 5:372. Translated as Immanuel Kant, "On a Discovery According to Which Any New Critique of Pure Reason Has Been Made Superfluous by an Earlier One," *The Kant-Eberhard Controversy*, ed. and trans. Henry Allison (Baltimore: Johns Hopkins University Press, 1973), 160.

to hold their views and to demonstrate how those views could be strengthened or weakened.

Since the heyday of analytic history of philosophy, a consensus has emerged that whatever the merits of the analytic historical approach for contemporary philosophizing, it had little to do with the history of philosophy itself. As historians of philosophy trained their sights on the historicity of history-of-philosophy discourse (tracing it back through a series of disciplinary and institutional reorientations in early nineteenth-century German universities and in some cases further), the notion that decontextualized accounts of pure conceptual structures were adequate to apprehend the historical meaning and development of philosophical discourse was rejected. The new emphasis reoriented the interpretation of philosophical texts (and particularly early modern ones) toward the project of understanding how the views and statements of individual philosophical authors would have been understood by those authors and their contemporaries. In most cases, the proponents of this view did not claim that such contextualism ought to displace what they saw as the "doing of philosophy."[11] Rather, they sought to establish a kind of meta-philosophical distinction between such 'doing' and its history, insisting that even the most seemingly atemporal and ahistorical of philosophical practices and views are still events or actions that take place within a historical context. This context, in turn, provides an interpretive lens through which those practices and views can be understood.

Often, such accounts began with a typology of approaches to work in the history of philosophy. Margaret Osler's more or less typical classification, for example, includes three modes: contextual history of philosophy, with attention to intellectual, social, cultural factors that "affect the way philosophers have thought about their subject;" analytic history of philosophy, which critically engages the ideas of historical philosophers as if they were participants in contemporary debates; and, though she does not name it as such, sophistical or ideological

[11] Popkin, 331.

history of philosophy, which uses "the ideas of historical figures to support their own philosophical agenda."[12] Within the first category, Osler distinguishes between those historians of philosophy who limit themselves to purely intellectual contextual elements (the class with which she identifies herself), and an "extreme version" of the discourse that emphasizes social and cultural history, something she explicitly identifies with Steven Shapin's sociology of knowledge and Foucaultian archaeology.[13] Overall, Osler argues for a contextualism focused on the agency and intentionality of philosophical authors, attacking positions that would attribute autonomy to ideas and concepts. Thus, the 'context' that interests her lies entirely within the bounds of the philosophical work, such that "At the most basic level, the context is the text itself."[14] By understanding the way philosophical texts offer evidence for the appropriation of ideas from the philosophical tradition and their conscious reconfiguration to deal with the philosophical problems concerning philosophical authors, the contextual historian of philosophy is able to grasp, insofar as possible, "the author's aim for the text as a whole."[15]

This emphasis on grasping authorial intention through contextual historicism has found, of course, its exceptions. Richard Watson, for example, points out in a Baylean vein that the history of philosophical misinterpretation is just as important — and in many cases more important — than an accurate narrative of authorial philosophical intention.[16] The pseudo-historical myths, fables, reconfigurations, and oversimplifications of the views of older thinkers are used to *deploy* a context against which 'newer' philosophers develop their own

---

[12] Osler, 529.

[13] Ibid.

[14] Ibid.

[15] Ibid., 530.

[16] Richard A. Watson, "Shadow History in Philosophy," *Journal of the History of Philosophy* 31:1 (Jan. 1993): 95.

theories and positions, and thus constitute what Watson calls "shadow histories of philosophy." His point is that these shadow histories are often just as important to understanding the history of philosophical discourse as the positions that the original figures actually held. What does it help us, a Watsonian historian of philosophy asks, to know that Bertrand Russell's influential 1900 reading of Leibniz contains many oversimplifications and distortions of Leibniz's actual views?[17] What matters is that this reading of Leibniz constructed a conceptual space for the development of Russell's own thought. This, the Watsonian insists, is more significant for understanding the last hundred years of philosophy than anything Leibniz himself ever wrote or meant to write.

The contextualists fire back that first, if we pause to consider the way that philosophers throughout the tradition have thought about their own mobilization of the history of philosophical discourse, such "shadow histories" have long been considered elements of the practice of the history of philosophy. Second and more significantly, emphasizing such shadow histories empties the magazine of the history of philosophy of its *historical* ammunition, such that a misinterpretation of Leibniz's late formulation of the principle of the identity of indiscernibles becomes equivalent to the factually erroneous claim that Leibniz was a teacher of Descartes.[18]

A more provocative criticism of the contextualist view comes from Thomas Lennon. While sharing the contextualist emphasis on the ineluctable historicity of philosophical discourse, Lennon remains suspicious of contextualist claims both about authorial agency or intention and the inviolability of the text as a given object of philosophical-historical interpretation.[19] While providing his

[17] For Russell's reading, see Bertrand Russell, *A Critical Exposition of the Philosophy of Leibniz* (London: Routledge, 1992).

[18] See Popkin's response to Watson in Richard Popkin, "Shadow History," *Journal of the History of Philosophy* 31:1 (Jan. 1993): 119-22.

[19] This discussion constitutes the final chapter (entitled "Philosophy and the Historiography of Philosophy") in Thomas Lennon, *The Battle of the Gods and Giants: The Legacies of Descartes and Gassendi, 1655-1715* (Princeton: Princeton University Press, 1993).

own typology of historical philosophical approaches, Lennon rejects the dichotomy between the history of philosophy and the history of ideas as traditionally construed in order to criticize them both. Both approaches are committed, Lennon argues, to a form (or really several forms) of assumed metaphysical realism. The history of ideas relies on a metaphysical realism in order to legitimate its principle of selection. The assumption here is simple: whatever the historian of ideas can interpret as a philosophical text must really be one. It follows (for the intellectual historian) that "philosophy is itself a metaphysically real natural kind," and thus that theories, ideas, and philosophical positions must be understood on the model of things.[20] Furthermore, the legitimacy of any historical account offered by this model depends on taking a realist view of history, which must be understood to be "metaphysically independent of anything that we might say about that past." This makes the critical issue of intellectual history the representational accuracy of an account relative to a really existing thing independent of that account.[21] Obviously, Lennon finds problematic this attribution of 'reality' to something that no longer exists.

Lennon also holds that the history of philosophy remains just as committed to a metaphysical realism as the history of ideas, even while the former aims for an atemporal and conceptual accuracy and the latter for one temporal and historical. If philosophers in radically different historical and social contexts can be understood to be discussing the same problems and views (the central thesis of Lennon's 'history of philosophy position'), then those problems and views must have some real status independent of their historical embeddedness and articulation. Thus, just as the history of ideas must hold that philosophy itself is metaphysically real, the history of philosophy must hold that "philosophical views

---

[20] Ibid., 379.

[21] Ibid.

are themselves natural kinds."[22] While rejecting the reality of history or at least relegating it to incidental status, this approach imbues philosophical positions, arguments, questions, and problems with more reality than both the agents who think and write about them and the texts in which they congeal.

Lennon ultimately proposes an historically informed and rigorous anti-realist approach to a history of philosophy that understands that history is to be produced through practices of interpretation. Unlike the history of ideas, an anti-realist history of philosophy, Lennon claims, would be able to deal with its own foundational metaphysical assumptions and principles of selection, since it is not forced to restrict itself to the historical reconstruction of the context of philosophical discourse.[23] Crucially, such an anti-realist view would have to take up a very different position toward the status of the 'evidence' or 'documentation' available for scrutiny. In their traditional (metaphysical realist) instantiations, both history of philosophy and history of ideas ground themselves on the reality and independence of the text taken as primary. The first does this in order to elucidate and evaluate the positions and concepts entombed within the text, while the second does this in order to reconstruct the contours of the past that informed the production of the text. Lennon proposes understanding the text instead to be a *product* of philosophical interpretation. While he does not explicitly cite the continental philosophical tradition here, it seems to me that Lennon's point is similar to one proponents of Gadamerian and Ricoeurean hermeneutics frequently make: the meaning of a text is the product of an act of interpretation that itself takes place within a specifically delimited historical and semantic horizon.[24] While the act of interpretation itself must take into account the historical conditions and context that give rise to the meaningfulness of discourse by

---

[22] Ibid.

[23] Ibid., 383.

[24] Although the alliance appears specific to the work, in his book on Pierre Bayle Lennon explicitly mobilizes Mikhail Bakhtin's dialogical theory. Thomas Lennon, *Reading Bayle* (Toronto: University of Toronto Press, 1999), 24-41.

seeking to understand the horizons of the world in which it *has* meaning, grasping that meaning is an interpretive act taking place within a determinate system of signification delimited by its own contextual horizons. Thus, historical philosophical interpretation cannot rely on the neutrality of the (metaphysically real) object of interpretation, but instead must understand itself to be playing a role in the instantiation of the text *as* a carrier or site of meaning.

I find Lennon's approach particularly important for the problem of this essay because he frames the problem of philosophical historiography in terms that bring it into contact with some of the theoretical approaches and methodological strategies dominant in other humanistic disciplines. While still recognizable as an historian who shares the contextualist emphasis on the historical embeddedness of philosophical discourse, Lennon remains suspicious of the metaphysical assumptions that the contextualists share with the analytic historians: philosophical ideas (now understood as the intentional products of autonomous actors within a determinate historical framework) are immaterial entities understood to be metaphysically real, and only accidentally embedded in more or less complex historical situations. Lennon, that is, opens the door to a critique of contextualism, but not in the service of its disavowal and the reestablishment of the independence of philosophical interpretation from history. Rather, his critique demands an expansion of contextualism both to the seemingly 'external' contextual factors (external, that is, to the agency of historical figures) that play into the production of philosophical discourse and to the interpretive horizons in which the contemporary act of historical-philosophical interpretation takes place. Both aspects of this critique, of course, were major loci for parallel theoretical shifts in the historicist turn discussed earlier with respect to the other humanistic disciplines.

One example of the implications of such an anti-realist critique can be found in the problem surrounding the establishment of the philosophical text itself. On a certain level, this is simply a question for philologists and editors of critical editions, whose task is to uncover the (metaphysically real) philosophical text obscured by the corruptions of history. Once a critical edition of a philosopher's work is produced and meets with general approval, we can agree to trust the editors to have made available the best possible printed instantiation of the material and to have elucidated any problems or questions surrounding the status of the text. There may even be good historical reasons for holding that questions of textual establishment are no longer as crucial for the philosophical work of early modernity as for earlier periods. Under the rubric of "typographical fixity," for example, Elizabeth Eisenstein argued that the early modern period is marked by the stabilization of print artifacts. Given the ubiquity of their availability, their reproducibility (infinite at least in principle), and the normalization of the conventions governing their format and appearance, seventeenth-century print culture allowed for forms of *dematerialization* whereby books and other printed matter became mere exemplars of a now inviolate content that reappeared on an identical page each time another printing was undertaken.[25] With printing presses having reached ubiquitous levels of dispersion in the seventeenth-century, and with formal juridical practices established to control the production and circulation of texts (such as the royal privilege, which functioned both as an early form of copyright and as a mechanism for censorship), for the first time readers could be assured that the philosophical text that they purchased from a local bookseller was the definitive text itself. We can recognize here an historical argument for the thesis of metaphysical realism: as manuscript and typographical singularity disappears, the reality of the philosophical text transcends its material instantiation.

More recently, however, book historians like Roger Chartier and Adrian Johns have challenged Eisenstein's assertions about the nature of early modern textual stability through analyses of the production of scientific and philosophical works. They point to the complex collaborations required among authors, editors,

[25] Elizabeth Eisenstein, *The Printing Press as an Agent of Change: Communications and Cultural Transformations in Early Modern Europe*, 1 vol. ed. (Cambridge: Cambridge University Press, 1979), 113-26.

printers (both legitimate and 'piractical'), booksellers, censors, and other state agents for the production of intellectual work.[26] Thus, while contextualist historians of philosophy depend on the claim that actual philosophical authors autonomously exercise their agency in the production of their texts, in some cases the 'product' available for scrutiny to the historian of philosophy is actually the result of a tension-fraught collaboration between authors, editors, printers, censors, and so on. It also follows from Chartier and Johns' analysis that the sheer quantitative explosion of written culture and the development of new series of forms for its circulation (for example, the correspondence networks, learned societies, and scholarly journals of the second half of the seventeenth-century) destabilize texts in at least two ways. These developments open a myriad of new avenues by which readers can approach texts and render more complex the chains of sovereign authorial production that connect authors to their texts and texts to their readers. I would add that, finally, the trend toward standardization of typography and printing conventions cited by Eisenstein serves, among other things, to draw *attention* to deviation from its norms, such that what might earlier have appeared to be a stylistic decision now allows—in the seventeenth-century—for the deliberate canalization of reading practices.

These weaknesses in the textual stabilization thesis allow us to see one reason why even thoroughly contextualist approaches to philosophical historiography tend to relegate historical specificity and materiality to the realm of *context*. Despite their undeniably positive moves, it still seems to me that the practitioners of these new forms of philosophical consideration still see these concerns as merely propaedeutic moves necessary for rigorously performing the *real* act of interpretation: a theoretico-philosophical reading of the theoretico-philosophical work, independent of all consideration of the role that the metaphysical assumptions standing behind such a reading play in the construction

---

[26] Roger Chartier, *The Order of Books: Readers, Authors, and Libraries in Europe between the Fourteenth and Eighteenth Centuries*, trans. Lydia G. Cochrane (Stanford: Stanford University Press, 1994). Adrian Johns, *The Nature of the Book: Print and Knowledge in the Making* (Chicago: University of Chicago Press, 1998).

of such a work. This privilege accorded to the interpretation of immaterial philosophical ideas—now seen as accidentally embodied in more or less complex historical situations—is, of course, one of the primary targets at which new historicism, microhistory, and critiques of contextualism like Lennon's take critical aim.

Leibniz's "Système nouveau" stands as an interesting philosophical example of the difficulties inherent in these metaphysical assumptions, including the sense in which they rely on the historical assumption of textual stability. Its status as a text-object is particularly complex, since there is a sense in which what philosophers refer to by this title continually slides away. We can point, of course, to its initial anonymous publication in the June 27 and July 4 1695 issues of the *Journal des savants*.[27] However, what historians of philosophy are generally pointing to when they discuss Leibniz's "Système nouveau" is a broader aggregation of texts, all of which follow, in various venues, the publication of these two journal issues: Simon Foucher's response in a September 1695 issue of the *Journal des savants*; Leibniz's so-called "First Explanation" in two April issues of the 1696 *Journal des savants*; the so-called "Second Explanation" published as a part of a section entitled "Extraits de diverses lettres" in the February 1696 *Histoire des ouvrages des savants*; the so-called "Third Explanation" (an extended version of the second) published in the November 1696 *Journal des savants*; Pierre Bayle's remark H to his "Jerome Rorarius" article in the 1697 *Dictionnaire historique et critique*, including substantial quotation from Leibniz's initial publication and references to Foucher's objections and Leibniz's explanations; Leibniz's letter regarding Bayle's remark in the July 1698 *Histoire des ouvrages des savants*; Bayle's remark L added in the 1702 second edition of the *Dictionnaire*, which reproduces portions of Leibniz's response to remark H; Leibniz's reply to remark L, published in the *Histoire*

[27] A version of the core text emended by Leibniz after publication is available in G. W. Leibniz, *Die Philosophische Schriften*, ed. C. J. Gerhardt, 7 vols. (Hildesheim: Olms, 1978), 4:477-87. For the text in its original state, consult *Le Journal des savants*, Paris ed., 25-26 (June 27/July 4 1695): 294-306 (http://gallica.bnf.fr/ark:/12148/bpt6k56542r).

critique de la république des lettres in 1716 after Bayle's death; the Amsterdam editions of all of these issues of the *Journal des savants* actually consulted and cited by Bayle; and finally, the pirated editions of the *Journal des savants* that appeared in Rotterdam earlier than the Amsterdam edition and were thus most likely Bayle's first point of contact with Leibniz's texts.[28]

The "Système nouveau" is thus a complex text-object with a myriad of material forms. If we want to understand what the "Système nouveau" means as a historical text, then we would have to know something about who was reading these journals and why, as well as how they went about assembling these fragments into a coherent whole.[29] We would also have to account for the complex printed format of Bayle's *Dictionnaire* and its curious apparatus, which sets up a visual and referential structure of relation between his own commentary, Leibniz's texts, work by recent interpreters of Aristotle, a fifteenth-century treatise on animal rationality, Pliny on the erotic and religious life of elephants, and so on. In other words, we need to think about the way that the "Système nouveau" is not a discrete text, but an assemblage of material objects that appear in a set of very particular locations, immersed in particular practices of reading, and provoking a variety of potentially incommensurable responses by seventeenth and twenty-first-century interpreters.[30]

---

[28] The Leibniz texts are collected in Leibniz, *Philosophische Schriften*, 4:487-503, 4:517-24, and 4:554-71. Most (but not all) are translated in *Leibniz's 'New System' and Associated Contemporary Texts*, ed. and trans. R. S. Woolhouse and Richard Francks (Oxford: Oxford University Press, 1997). Pierre Bayle, *Dictionnaire historique et critique*, 5th ed., 4 vols. (Amsterdam: P. Brunel et. al., 1740), 4:82-83 and 4:85-87. I mention the pirated editions of the *Journal des savants* because the intense demand required to produce a black market edition suggests a different sort of readership for the journal than might otherwise be assumed. See Jean-Pierre Vittu: "Le contrefaçons du *Journal des savants* de 1665 à 1714" in *Les Presses grises: La Contrefaçon du livre*, ed. François Moureau (Paris: Aux amateurs de livres, 1988), 303-23.

[29] On the changing readership of the intellectual journals, see Jonathan Israel: *Radical Enlightenment: Philosophy and the Making of Modernity, 1650-1750* (Oxford: Oxford University Press, 2001), 142-155. On the initial readership of the *Journal des savants*, see Harcourt Brown: "History and the Learned Journal," *Journal of the History of Ideas* 33:3 (Jul.-Sept. 1972): 365-78.

[30] My purpose here is not to carry out these tasks, but rather to frame the kind of investigations that it would take to offer the kind of account of the "Système nouveau" that would satisfy conditions for an investigation of it under the rubric of a material philosophical historiography.

How incommensurable those responses might be can be judged in part by the set of extremely diverse interests of the newly developing community of readers—the Republic of Letters—in which Leibniz's text circulated. Consider, for example, the articles and reviews surrounding the initial publications of the "Système nouveau," Foucher's reply, and the first and third explanations. A list of these includes: a refutation of Quietism, a new philosophical dictionary, the rules of a particular monastic discipline, a pharmaceutical collection, a method for learning to read in only a few days, a technical manual for constructing a machine to remove water from foundations, reports on the proceedings of recent ecclesiastical conferences, a book on the passions, a *Traite de la satire* examining both its polemical use and its future, a book entitled *Réflexions sur l'élégance et la politesse du style* (admittedly a review of a treatise on orthography rather than a Routledge-like cultural history of early modern French fashion), a reprint of a letter originally published in an English journal detailing (with diagrams) the dissection of a prostitute's reproductive organs (that is, early modern intellectual pornography), a treatise on practical geometry for use in the construction of fortifications, a biography of a prominent churchman, a review of a new book on the history of the eruptions of Mount Vesuvius, an explanation of the Ash Wednesday ceremonies, a philological study of a recently discovered manuscript by St. Gregory, a treatise entitled *Les Éléments de l'histoire*, a biography of the Queen of England, and a study of the anatomy of the brain.

While it would certainly be absurd to say that all readers interested in Leibniz's texts would have been interested in all of these topics, they would certainly have been reading the journals for more than just philosophical articles like Leibniz's. On the other hand, we need to remember that Leibniz's ideas in the "Système nouveau" would have reached and been at least somewhat digested by an audience fascinated by the construction of fortresses, literary theory, the female reproductive system, the everyday life of monks, the history of volcanoes, and the Queen of England. The "Système nouveau," therefore, is an extremely difficult text to even identify discretely *as* a text, and an account of its circulation, reception, and interpretation requires a history of seventeenth-century journal and epistolary culture, the sometimes paradoxical social and intellectual roles played by Bayle's *Dictionnaire*, its complex visual format, and so on. If any philosophical text can be said to exceed or to problematize its limits, then surely this is one.

I will now briefly outline Leibniz's account of the phenomenality of material objects as presented in the "Système nouveau" and suggest that it might be used to formulate a solution to the problem of dealing with the material reality of the philosophical text without forsaking its intellectual content or confusing that material reality with the metaphysical reality of the ideas it expresses. I will not address the vast majority of philosophical issues in the "Système nouveau;" my purpose is simply to interpret the material-historical status of the set of texts we call the "Système nouveau" on the basis of the theory of materiality it contains. If nothing else, understanding this text in terms of the doubled reference of conceptual content to historical status and vice versa will point to one way out of the history-theory impasse as outlined above.

Though the story becomes extremely complicated when the diversity of mechanical, corpuscular, and atomistic schools are taken into account, I think it is fair to say that the dominant continental intellectual school of the third quarter of seventeenth-century philosophy (Cartesianism) understood materiality in terms of extended substance. Bodies, that is, were defined essentially by their dimensionality. Corpuscularists arguing that the parts of matter constituted natural *minima* (instead of possessing the infinite divisibility Descartes proposed) challenged elements of this stance, of course. Both schools, however, agreed that nature must be understood as a mechanical system of matter in motion, with that matter possessing (and for the Cartesians, being defined by) length, depth, and breadth. When he steps into the debate over the nature of matter, Leibniz replaces extension with force as the proper measure of a body, because, he claims, the discrete edges of objects are illusions produced by our inability to perceive the folds of their infinitely divided matter. Extension, Leibniz holds, is only a matter of phenomenal contact and involves no "substantial unity."[31] He also rejects Descartes' two modalities of extension — infinite divisibility and determinate shape. Descartes insisted that everything in the corporeal world was mechanically constructed from infinitely divisible corpuscles of matter, each part of which could be divided in any way we please.[32] For Leibniz, the infinite divisibility of matter is not merely potential and modal but actual: the material world *is* divided, not merely divisible. Material bodies are "natural machines," "machines with a truly infinite number of organs," "machines composed of machines even in their smallest parts."[33]

In the "Système nouveau," Leibniz offers an account of his rejection of atomism on the basis of what he would elsewhere call "the labyrinth of the continuity of the continuum," here framed in terms of that which is capable of constituting the reality of multiplicity (la multitude). Put simply, Leibniz's claim is that the foundational unity through which multiplicity can be grounded cannot be found in the material dispersion of atomistic particles or quasi-atomistic insensible bodies. Insofar as something non-metaphysically simple—a multiplicity, or a complex body—is real, its reality must derive from something other than its constituent parts. Thus if we observe the world around us and note that everything appears to exist within a continuum, we will have to seek the reality of that continuum somewhere other than in the material out of which it seems to be composed. The problem is that the reality of what Leibniz calls "the principles of a true unity" cannot be found in matter alone, since matter, in his

---

[31] Leibniz, *Philosophische Schriften*, 4:482; *Leibniz's 'New System'*, 16.

[32] The key texts here are parts of the *Principia philosophiae* and *Le Monde*, in René Descartes, *Oeuvres de Descartes*, ed. Charles Adam and Paul Tannery, 11 vols. (Paris: Vrin, 1996), 8:40-79 and 11:31-48.

[33] Leibniz, *Philosophische Schriften*, 4:480-82; *Leibniz's 'New System'*, 16-17.

view, "is nothing but a collection or heap of parts to infinity" that, in terms of agency, is merely passive.[34]

In order to establish the reality of composite or aggregational things, Leibniz argues that they must 'contain' something other than the mass of material parts that stand in a particular configuration. The reality of an army cannot be explained on the basis of the disposition of a determinate number of soldiers and weapons alone; the reality of a pond full of fish is not due to fish bodies swimming in various configurations; the reality of a flock of sheep is not due to a set number of animal bodies huddling together; and the reality of a watch is something more than the set of relations in which its gears, springs, face, and hands stand. Insofar as we restrict our account of purely material configurations to their constituent parts, Leibniz claims that they are not, strictly speaking real. Understood in this sense, the soldiers may be real, but the army is not; the fish and water molecules may be real, but the pond is not; the sheep may be real but the flock is not; the gears and springs may be real, but the watch is not.[35] Only by introducing what he calls in various places within the "Système nouveau" "real unities," "formal atoms" or "atoms of substance," "real and animated points," "substantial forms," and "primary forces" (all of which are equivalent terms at this point in Leibniz's intellectual development) can we ensure the reality of the multiple. Material atomistic bodies, he claims, are themselves composed of parts, even if these parts are indivisible from the bodies themselves.[36] Despite this indivisibility, Leibniz holds, these parts are still not identical to one another, and their metaphysical difference precludes their functioning as "absolute first principles of the composition of things."[37] Having thus dispensed with atomistic points and replaced them with metaphysical points, Leibniz also considered the

---

[34] Ibid., 4:478; translation mine.

[35] Ibid., 4:482; *Leibniz's 'New System'*, 16. These examples are Leibniz's own.

[36] Ibid.

[37] Ibid.

nature of mathematical points. These, he claimed, "are the points of view through which metaphysical points express the universe." They are "exact," that is, indivisible limits to the extension of material things, the elements used to map the edges of materiality. At the same time, however, they are merely ideal and thus unreal: they are mere "modalities" of substances, the latter finding the principle of their reality in metaphysical rather than mathematical points.[38]

Even while the account of materiality presented by the "Système nouveau" is not necessarily the end toward which it aims, it may be useful for returning to the broader issue of how to conceive the materiality of a philosophical text, or more broadly, how to read philosophy from a material historicist perspective. So what, exactly, is Leibniz telling us about the nature of material things? Given Leibniz's stance that matter is actually divided and not merely infinitely divisible, and given that the edges of extended bodies can be defined by exact mathematical points that are themselves ideal, the edges and limits of material things, strictly speaking, are indeterminate and non-existent. For Leibniz, substantial unity requires immaterial principles of perception, interpretation, and force, the "metaphysical points" for which bodies are perceptual phenomena indicating that region of corporeality that a mind expresses distinctly and intensely. The *true* boundaries of bodies leak out imperceptibly to encompass the whole of the corporeal world. Material things, that is, are relatively amorphous *regional phenomena* stretching themselves into the heart of matter, non-localizable in any finite and particular place, bound by no truly distinct shape, manifesting only where they congeal a locus of perception.

If it is legitimate to turn this account of materiality back on the problem of the material and historical status of the "Système nouveau," then we get an interesting result. We are no longer faced with the quandary of delimiting the text as a material object. Materiality is now by Leibnizian definition a non-delimitable and ideal quality. As a criterion for textual interpretation, however, it has not simply disappeared in favor of an attempt to account for the intentions of its

---

[38] Ibid., 4:482-3; translation mine.

vanished author. Instead, that resistance to delimitation can become the interpretive principle by which we give an account both of the 'Leibnizian' position and of the conditions for its articulation. This approach is amenable to the more positive elements of the contextualist view without succumbing to its assumptions regarding metaphysical realism. One could even, I think, reflexively legitimate a traditional contextualist reading out of the account that the "Système nouveau" offers of "substantial atoms" (as I have tried to do here with its account of materiality), where the defining features that constitutes the *unity* of the text serve as the lynchpin for interpretation, rather than its dispersion as a material and historical text-object.

The text we are faced with when examining the "Système nouveau" is an aggregate of other texts, but then, every body is an infinite aggregate of other bodies. As a material — and by extension, historical — object, the "Système nouveau" should be understood as a text that spills over its phenomenal borders into the broader historical (and microhistorical) frameworks in which it circulated. The history of its multiple instantiation in the journal culture of the Republic of Letters and the complex characters and expectations of those journals' readers are no longer merely adjunct and contextual element of its interpretation, but part of the 'text itself.' The fact that most readers discovered Leibniz's text in Bayle's *Dictionnaire* — which functioned as both the epitome of late-seventeenth-century learned discourse and as its most popular fetishized commodity — is no longer something external to our 'reading' of Leibniz. Bayle's textual framework and the way it brings Leibniz's words into oblique contact with seemingly unrelated texts and demanded a particular practice of disjointed reading is something with *both* historical *and* philosophical significance. From the other direction, approaching the "Système nouveau" as a singular textual event with a particular material history is no longer a barrier to engaging its (now impure) conceptual content. We can do so without beginning with a disavowal of theoretical engagement, since such engagement is a necessary step in formulating our interpretation of its historical materiality.

Does this example of an apparent possibility for reconciling materialist historicism with philosophical-conceptual interpretation contribute in a significant way to overcoming their impasse? Does it truly begin to map a methodological and theoretical terrain that can address the commitments of historicism on the one hand and those of philosophical interpretation on the other? Probably it cannot. The problem of the methodological impasse between history and philosophy is not about to disappear. Creative scholarly debate over the possibilities for its solution or dissolution is essential if competing disciplinary models are not to simply continue talking past one another. If nothing else, my hope is that this essay contributes a productive (if problematic) example to that discussion.

Daniel Selcer (Duquesne University) is Assistant Professor in the Department of Philosophy at Duquesne University. He is completing two projects, one on the mobilization and appropriation of figures of printed books, encyclopedias, and libraries in the continental philosophical discourse of the seventeenth century; the other on Gilles Deleuze's reading of early modern philosophy.

The Holy Laws of Nature:
Antonio Vallisneri, Antediluvian Men, and the Flood

Michael Cunningham

Introduction

Antonio Vallisneri, a naturalist and professor of medicine at the University of Padua from the years 1700 to 1730, made frequent summer expeditions into the hills and mountains of northern Italy during which he observed a great many fossils. Most of these fossils resembled living marine creatures, yet he found them buried in rock formations high up on the hillsides many miles from the sea, which was a puzzle that sparked his interest. Because it touched on some of the foundations of natural philosophy, this puzzle was not only of interest to Vallisneri but also to naturalists, natural philosophers, and even theologians throughout Europe. Had God placed them there in the original creation? Had they grown directly in the rock because of its inherent "plastic virtue," or were they the remains of once living sea creatures? Did the biblical Deluge have a role to play in their present location far from the sea, or had there been a more natural process at work? The answers to the riddle of the fossils, then, had a direct impact on contemporary beliefs about God's role in the universe, humanity's origins, history, and natural law.

By the turn of the eighteenth century, naturalists' acceptance of the Mosaic account of creation, the universal Deluge, and the providential nature of change was slowly being undercut by critical readings of Genesis informed by new ideas of natural law. Were creation and the Deluge miracles, and thus outside of natural law, or did they follow natural law just as do all observed physical phenomena? Some of the widely held theories about fossil formation, the origin of the rivers, and the mutability of the Earth's surface, many of which were derived from

Aristotle's natural philosophy, came under increased scrutiny. The end result was a dramatic shift in humanity's place in *time*; instead of a comfortable six millennia of history, naturalists began to think that the Earth's history might be many thousands, and perhaps tens of thousands, of years. As was the case in the switch from geocentric to heliocentric cosmology, this shift in natural history was very unsettling to natural philosophers, and the long-range metaphysical consequences were to undercut the philosophical traditions founded on the synthesis between Christian theology and Aristotelian natural philosophy.[1]

Although the Aristotelian-Christian synthesis was under attack in the late seventeenth century, it was still the predominant mode of thinking, and there was a great effort to reconcile Genesis with the latest trends in natural philosophy, physics, and mechanics. Naturalists like Thomas Burnet and John Woodward developed theories about the origins of the Earth that were consistent with both mechanical law and the Mosaic account of creation.[2] Burnet's treatise covered the entire history of the world, from creation to eventual destruction, and did so in a way that followed the Biblical story in a general sense but it did not strictly adhere to every detail. Burnet's work did not directly address the origin of the fossils, but Woodward quickly stepped into the breach. Woodward embraced the new theory that fossils were organic in origin and argued that fossils had been created in the antediluvian era when the earth was covered in water, and then moved to their present locations by the cataclysm of the Flood. It is fascinating to note that Burnet's and Woodward's "sacred histories" were in some ways counterproductive; in using natural law to account for events previously held to be

---

[1] A good general introduction to early modern natural history is Paolo Rossi, *The Dark Abyss of Time: The History of the Earth & the History of Nations from Hooke to Vico*, trans. Lydia G. Cochraine (Chicago: University of Chicago Press, 1984). On the debate about the origin of the fossils, see Martin J. S. Rudwick, *The Meaning of Fossils: Episodes in the History of Paleontology* (New York: Neal Watson, 1976).

[2] Thomas Burnet, *The Sacred Theory of the Earth* (London: R. Norton, 1684); John Woodward, *An Essay toward a Natural History of the Earth* (London: Richard Wilkin, 1695).

providential, and thus supernatural, God's role becomes much more distant, opening the door for naturalists like Vallisneri to examine it more critically.

In 1721 Vallisneri published *De' Corpi marini che si Monti si trovano* (*On marine Bodies found on the Mountains*), his signature work on natural history, which covers in detail his opinions about the origins of the fossils, the universality of the Deluge, and the question of why antediluvian humans lived so much longer than their modern counterparts.[3] *De' Corpi marini* is a collection of fourteen letters, which Vallisneri put together very quickly in the summer of 1720. The introductory letter was one he received from Sebastiano Rotario, entitled "A Description of various crustaceans, and products of the Sea, that are found on the Mountains of Verona," wherein he described the strata in which Rotario found mummified insects and fossilized shells of crustaceans. The "First Letter" is Vallisneri's reply, simply entitled "Response of Mr. Vallisneri to Mr. Rotario," which contains his theory of fossil formation and some of his thoughts about the universality of the biblical Flood. The next letter in the book is also Vallisneri's, entitled "Second letter of Mr. Vallisneri to Mr. Abbot Lioni, on the Marine productions that are found on the Mountains, and the effects of the Deluge and the long lives of the antediluvian men." In the book, Vallisneri argued that fossils were organic in origin, deposited in an antediluvian epoch before the Biblical Flood when the northern Italian plain was covered by water. This water later receded, leaving the fossils high and dry on the hillsides of the Apennines. The natural inundations of which Vallisneri wrote would take time and he believed that the earth was older than the six thousand years held by traditional thinkers. The "Second Letter" addresses the question of why humans living before the Biblical Deluge enjoyed such long lives. Vallisneri explained that the Flood had a damaging effect on the food supply and the overall quality of the environment, which in combination with the psychological stress of the cataclysm, had an

---

[3] Antonio Vallisneri, *De' Corpi marini che si Monti si trovano* (Venice: Lovisa, 1721).

adverse impact on the human reproductive system and thus severely reduced the average lifespan. Vallisneri's discussion of fossil formation, the Flood, and the antediluvians, then, was focused on natural causes rather than divine intervention.

Yet in the "Second Letter" Vallisneri also argued that the Flood was "miraculous," posing the question of Vallisneri's attitude toward religion. Did Vallisneri concede that the Flood was miraculous only to make the "necessary protestations" – as he called them – in order to satisfy the Catholic censors? There is ample evidence to support this argument, and, judging by some of his comments in his correspondence, it does seem that the pervasive culture of Catholic orthodoxy in Italy forced him to dissimulate in his published works. Yet there is also some evidence to show that Vallisneri's acceptance of the long lives of the antediluvians may have been consistent with his general system of natural history. Orthodox natural historians had a difficult time fitting the antediluvians' nine-hundred-year life spans into their six-thousand-year timeline of Earth's history, so they treated the Biblical account, usually taken literally, as fabulous on this point. Since Vallisneri's conception of natural history, and his theory of fossil formation, required a great deal of time, the extended lives of the antediluvians fit quite easily into his universe. In the final analysis, Vallisneri is best described as a Catholic in the Galilean tradition, who believed that matters of faith and natural philosophy should be separate.

The Origins of the Earth

Natural history in the early modern period, and especially the late seventeenth and early eighteenth centuries, addressed some of the most fundamental questions about the universe. Where did we come from? How was the Earth created? What is our place in the world? These questions had been answered authoritatively by orthodox Christianity – with some slight variations between Protestant and Catholic – based on the biblical tradition. Naturalistic answers to these questions based on empirical evidence and the geological record were, due to the tremendous impact they had on matters pertaining to theology and natural philosophy, hotly contested. If the Mosaic account of creation were to be proven incorrect using the new methods of natural philosophy, then the foundation upon which Christian morality rested would be weakened and the whole edifice might come crashing down. Moreover, Europeans were beginning to apply natural law to human society, which seemed to support existing social structures and reinforce the native superiority of the noble families of Europe.[4] Thus in the debates about the origins of the earth, the very foundations of early modern society seemed to be at stake, and it is not surprising that the question attracted some of the greatest minds of the seventeenth century, including René Descartes, Sir Isaac Newton, and Gottfried Wilhelm Leibniz.[5]

Descartes posited perhaps the most controversial theory of Earth's origins, one based on the mechanical philosophy. Mechanism asserts that matter is composed of indivisible particles called atoms, that all matter has motion, and that everything in nature is connected via a universal plenum. Since the universe is completely filled with atoms, there is no allowance for a vacuum in the mechanical philosophy. According to mechanism, the world is like a machine that follows natural, mechanical laws. Newton's mysterious attractive force of gravity was not possible in the Cartesian cosmos; any action at a distance must be the result of a direct, material connection between the two objects. Descartes was very careful to describe the importance of God in creation as a kind of divine Clockmaker, but in his *Principia* he described a hypothetical version of creation whereby the world gradually evolved from fiery particles into denser matter

---

[4] For the discussion in the Royal Society about this subject, see Rhoda Rappaport, *When Geologists Were Historians* (Ithaca, NY: Cornell University Press, 1997), 87-89 and Steven Shapin, "Of Gods and Kings: Natural Philosophy and Politics in the Leibniz-Clarke Disputes," *Isis* 72 (1981) 215.

[5] On Descartes, Newton, and Leibniz, see Rossi, 44-54.

through the constant motion of its atoms. Descartes had a number of supporters in France and Italy, but very few in England.[6]

Isaac Newton and Robert Boyle, who both embraced the natural law but denied its relevance to the creation of the universe, led the English reaction against Descartes. In their view, only Epicureans and atheists, terms that were synonymous in their usage, would attempt to apply mechanistic principles to the first origins of the Earth. The Newtonian view of matter supposes that it has no inherent motion and so mechanical principles alone could not provide the basis for creation – only the power of a divine being could sufficiently account for it. According to Newton, God had created the universe out of chaos in seven days, probably much longer than our present days, and then set it to work according to natural laws. In Newton's view, the Mosaic account was the most important text for understanding the Earth's origins, and any attempt to analyze the events of the providential and miraculous creation using reason or natural law was, by definition, useless.

Taking a middle position was Leibniz, who believed that God chose our particular world from all of the possible worlds, that matter appears only in the forms compatible with this world, and that there has never been chaos. The chaos that is apparent to human observers exists because we cannot perceive the order inherent in the most basic forms of matter. As Liebniz wrote, "In this way, the apparent Chaos is but a sort of distance: as in a pond full of fish, or better, as in an army viewed from afar, from where it is impossible to distinguish the order observed there."[7] God had chosen the world, and it operated according to natural laws that may be understood by humankind, but only if observed from the proper distance. For Leibniz, the Earth has never been a paradise; from the imperfect

---

[6] Most notably Fontenelle in France.

[7]Gottfried Wilhelm Leibniz, *Die philosophischen Schriften*, vol. 3 (Berlin: Herausgegeben, 1875-1890), 565, as quoted in Rossi, 54.

human perspective there has always been disorder. Concerning the mechanics of formation, Leibniz believed that the Earth coalesced from a globe of very hot liquid matter. As it cooled it formed a crust and water was thrown into the atmosphere by exhalations of humid vapor. Because all matter does not cool at the same rate, blisters appeared on the crust that became mountains and valleys. At the very beginning of the Earth's history, the globe was mostly covered by water, which then slowly receded into subterranean caverns and exposed more land area. It was at this time that the fossils were created; he believed that they were imprints of real fish left behind on dry land as the waters receded. According to Leibniz, once the basic structure of the Earth was in place, there were no further *general* mutations; all future changes were *particular* in nature. The Flood is a good example of this – it was a discreet discontinuity in the Earth's development and something that had particular consequences.[8] Liebniz's view was extremely optimistic, a characteristic Voltaire satirized mercilessly in *Candide*, but it had the advantage of allowing the universe to be both chosen by God *and* created according to universal natural law.

Vallisneri's position on the Earth's origins shared more with Leibniz's than Newton's or Descartes' theories. Vallisneri corresponded in Italian with Louis Bourguet, a philosopher and mathematician at the University of Neuchatel, and Leibniz wrote to Bourguet in French, so Bourguet acted both as intermediary and contributor in an ongoing three-way dialogue. Perhaps Leibniz's most important contribution to Vallisneri's work was the idea that the Earth was originally covered in water, which slowly receded over time – a theory Vallisneri adopted in his work on the origin of fossils. Vallisneri agreed with Leibniz's argument that the Flood did not destroy the foundations of the mountains, as naturalists like Burnet and Woodward posited, nor did it completely destroy the

[8] See Gottfried Wilhlem Leibniz, *Protogaea* (Goettingen: C. L. Scheidio, 1749); also Rossi, Chapter 9.

surface of the world. Vallisneri was not, however, an uncritical follower of Leibniz. In a letter to Bourguet dated September of 1720, Vallisneri wrote, "the most probable [theory of the Earth's origins] is that of Leibniz, but his also has large difficulties."[9] Vallisneri is similarly cautious in *De' Corpi marini*: "I ask, if the ingenious opinions of Leibniz's – that these caverns under the ground were never completely full – were true, would not the abundant water above enter them?"[10] Vallisneri's relationship with Leibniz was thus highly creative and mutually productive.[11]

Following the lead of Leibniz and Newton, most late seventeenth-century naturalists retained the biblical account of creation as the foundation for their philosophy, but it was a foundation that was becoming progressively weaker and some turned to natural forces as a bulwark. The weakness was caused by new forms of criticism that treated the Bible as a historical document rather than a font of received knowledge. Baruch Spinoza, for example, in his *Tractatus theologico-politicus* of 1670, put it under the same kind of critical lens as historians had done with Herodotus, Thucydides, and the Roman chroniclers.[12]

---

[9] Vallisneri to Bourguet, 18 August 1719, BVN, *Ms.* 1282, letter 231.

[10] Vallisneri, *De' Corpi marini*, 91. One finds many references to Leibniz throughout Vallisneri's writings, from his ideas on the geological strata in *l'Origine delle Fontane* to his analysis of fossils in *De' Corpi marini*. Vallisneri directly acknowledged both Leibniz and Girolamo Fracastoro on pages 32-47 of *De' Corpi marini* as he discusses the origin of the fossils. It is not useful to say that Vallisneri, nor any other Italian scholar, was a "Leibnizian" in the same way that many historians use "Galilean" as a definitive category. But if Vallisneri was not a true disciple of Leibniz's, his preference for Leibniz's, rather then Newton's or Descartes's, more general philosophies illustrates the dangers of arguing, as Vincenzo Ferrone does in *The Intellectual Roots of the Italian Enlightenment*, that an acceptance of Newtonian ideas was the foundation of modern scientific thought in Italy. Newton's philosophy did not have the kind of privileged position that many modern historians ascribe; a teleological approach may tempt one to look for the origins of modern science in the spread of Newtonianism, but in northern Italy Leibniz had at least as much influence.

[11] Leibniz to Bourguet, 22 March 1714, in Leibniz, *Die philosophischen Schriften*, vol. 3, 565.

[12] Rappaport, 74.

Because of the problems inherent in translating ancient languages, he insisted that one could not know with absolute certainty the precise meaning of many biblical passages. Received knowledge was slowly losing its authoritative status, and, rather than using the Bible as a guide to understanding nature, natural historians began to use nature as a guide to understanding the Bible.[13] The new field of inquiry was very appropriately called physico-theology, or natural theology, and it attracted a great deal of thought in the later half of the seventeenth century.

Considering the Roman Catholic Church's commitment to maintaining its authority in the realm of natural philosophy, and the Protestants' relatively freer stance on astronomy and physics, as evidenced in the trial of Galileo in 1633, one might expect that some of staunchest defenders of the biblical account of creation would be Catholic, but this was not the case. With the Reformation, Protestants had rejected the authority that had been vested in the Catholic Church since the medieval era, but had replaced it with a reliance on the literal word of God as it had been recorded in the Bible. For Protestants, perhaps even more than Catholics, the words written in the Bible were absolute truth. So it was especially important for Protestants to reconcile the empirical evidence found in nature with the mythopoeic evidence found in Genesis.

One of the events in Genesis that was of great interest to naturalists in their quest to explain the world's history was the universal Flood. At the turn of the eighteenth century most naturalists agreed that the biblical Flood actually happened, but there was much disagreement about whether it was caused by supernatural or natural forces. Supporting evidence for the Flood as a natural event included Holy Scripture, Sumerian mythology as written in the *Epic of Gilgamesh*, and empirical evidence taken from observations of 'erratics' that had

---

[13] Ibid., 82.

been deposited by glaciers that had once covered northern Europe.[14] Naturalists like Thomas Burnet and John Woodward attempted to reconcile Scripture with natural law, explaining the origin of the earth, the Flood, and the origin of the fossils on natural terms. Others, like Vallisneri, avoided the issue by describing the Flood as a miracle, thus keeping it outside the realm of natural history. Pious naturalists, in particular, were concerned about natural explanations of the Flood; if it were not the universal cataclysm of divine punishment described by the Bible, but rather a natural event with no moral purpose, the authority of Christianity, and with it the established social and political order, could be called into question. Because of its great importance, the subject inspired many passionate and sometimes polemical works in the late seventeenth century on all sides of the issue.

A Sacred Theory of the Earth

Thomas Burnet's *Telluris theoria sacra* (1680), published as *Sacred Theory of the Earth* in 1684, was the first major work to use natural forces to explain biblical events, and it set the tone for future discussions. Because he saw nature as the creation of God, Burnet saw no inevitable conflict between religion and reason; an inquiry into nature was an inquiry into the mind of God. Burnet sought to maintain a balance between reason and the holy word, to discover a kind of natural providence. One of his main goals was to protect the doctrines of the universal Deluge and paradise "from the Cavils of those who are no well-wishers to Sacred History."[15] The result was a 'sacred theory' that was a fusion of text-based traditional scholarship and Cartesian analytical thinking.[16] Burnet's

---

[14] David R. Oldroyd, *Thinking About the Earth: A History of Ideas in Geology* (Cambridge, MA: Harvard University Press, 1996), 50.

[15] Burnet, a3.

[16] Rudwick, 77.

conception of the origin of the earth was especially creative: he argued that it had coalesced from water into a perfectly smooth body that was transformed into its present irregular form during the cataclysm of the Deluge. His account of creation relied heavily on Genesis, especially the passage concerning God's power moving over the water. "And the earth was without form, and void; and darkness was upon the face of the deep. And the Spirit of God moved upon the face of the waters."[17] Burnet took this description and applied a kind of scientific analogy to the text. He pointed out that liquid is unstructured matter, essentially matching the definition of chaos, and so one can take the reference to waters literally and say that from them (a chaotic liquid mass) God created the Earth. Burnet's God did not contravene natural law with creation; the process was much like ice formation on a lake. When water congeals it creates a relatively flat solid surface, and so, by analogy, Burnet argued that the original surface of the Earth was like a billiard ball, smooth and featureless with no seas, mountains, valleys, or rivers.[18] Continuing the analogy, Burnet observed that underneath the ice on a frozen lake is unfrozen water, and the antediluvian Earth, too, had a great volume of water underneath the solid crust. The Great Abyss that appears in the Bible, then, must be this mass of water. Supporting the Abyss was the solid center of the Earth, a perfect sphere of congealed metal formed from the pre-creation liquid chaos. Burnet's antediluvian Earth, then, had three parts: a smooth outer crust, a layer of water, and a solid metal core.

Burnet's discussion of the creation was clearly aimed at refuting the Aristotelian position that the world is eternal, and he used a combination of reason and textual analysis to prove his point. "Having this infallible assurance of the Origin of the Earth and Mankind, from Scripture, we proceed to refute the same

---

[17] Genesis, 1:2. All quotations are taken from the King James version of the Bible.

[18] Burnet, 50.

Doctrine of Aristotle's by natural Reason."[19] The mechanistic analogy of the universe as a watch was useful to him in this task; an eternal watch was impossible, he argued, because at some point the parts had to be put together. Moreover, considering the power of the natural forces of erosion and volcanism, it was unimaginable that the Earth could remain the same forever. Why have the mountains not been reduced to plains by the constant action of wind and rain?[20] In addition to the Bible, Burnet also used human history and his knowledge of the classical authors to support his argument. If, as Aristotle had claimed, the world had existed forever, why did the ancients not know more about the world's geography? If they had an infinite amount of time to explore the Earth's surface, they should have known more than the area around the Mediterranean. Having disputed Aristotle, then, Burnet was content that he had established a sacred theory of the world's origins and saved it from the threat of atheism.

Burnet's attempt to use natural law to explain the process of creation may have seemed like a good strategy in the fight against atheism, but it was inherently risky because it could be interpreted as an attack on the literal truth of the Mosaic account and, by association, the fundamentals of Christian theology. Burnet's readings of Genesis were often metaphorical and highly creative, and his descriptions of biblical events often seem to be primarily natural rather than supernatural. For a philosopher interested in promoting sacred history, he was almost Galilean in his approach to biblical exegesis. "'Tis a dangerous thing to ingage in the authority of the Scripture in a dispute about the Natural World, in opposition to Reason," he wrote, "lest time, which brings all things to light, should discover that to be evidently false which we had made Scripture to

---

[19] Ibid., 35.

[20] Ibid., 37.

assert."[21] Some of Burnet's contemporaries, believing that it opened the door for atheism, criticized him for using natural principles to explain inherently miraculous events.[22] This, of course, was far from Burnet's intention, but his critics had a point; his theories did open some space for naturalistic accounts of creation.

The Great Protector of the Universal Deluge

One of the naturalists who attempted to fill in the gaps in Burnet's vision was John Woodward, whom Vallisneri called "the great Protector of the universal Deluge."[23] In his *An Essay Toward a Natural History of the Earth* (1695), Woodward argued that the biblical Flood played a critical role in fossil placement, explaining why fossilized marine animals could be found on mountaintops far from the sea. Woodward's work touched off a general discussion of fossil placement in both England and the Continent, to which John Ray, William Whiston, Louis Bourguet, Vallisneri, J. J. Scheuchzer, and many others contributed.[24] Although some naturalists objected to Woodward's specific points about how the Flood distributed the fossils, it is safe to say that most of them were diluvialists who believed that the Flood was universal and had played a major role

---

[21] Ibid., a.

[22] Rappaport, 139-149. Vallisneri took this attitude in *De' corpi Marini*, see especially the "Second Letter." Another example is Anton Lazzaro Moro, who wrote in 1740 that the Deluge could only be understood as "one of the most solemn prodigies of Omnipotence," and that it was "prodigious extavagance" to explain it with the laws of hydrostatics. Anton Lazzaro Moro, *De' crostaci e degli altri corpi marini che si truovano su' monti* (Venice, 1740). Translation from Rossi, 78.

[23] Vallisneri's comment about Woodward can be found in Vallisneri, *De' Corpi marini*, p. 26.

[24] On the reception of Woodward's work, see Rappaport, 149-157. Most of this discussion took place in letters and articles, except for Whiston's book. William Whiston, *A new theory of the earth...* (London: R. Roberts, 1696). In 1720 Vallisneri had not yet read Whiston's work, so it had little impact on Vallisneri's *De' corpi Marini*.

in fossil placement. *An Essay Toward a Natural History of the Earth* was still the seminal work in the field in 1720 when Vallisneri was collecting for publication his arguments about fossils.

According to Woodward, fossils had originally formed in the earliest stages of the earth's development, after the divine act of creation, when it was still covered in a great deal of water. Obviously the earth's surface had changed since that era, so there had to have been some "alterations" to explain the current location of the fossils. In Woodward's conception, everyday forces of nature could not have produced such a jumble of stone, marble, and chalk; only the Deluge had enough power to completely reconfigure the Earth's surface. During "the time of the Deluge, whilst the Water was out upon, and covered the Terrestrial Globe, All the Stone and Marble of the Antediluvian Earth: all the Metalls of it: all Mineral Concretions: and, in a word, all Fossils whatever that had before obtained any Solidity, were totally dissolved, and their constituent Corpuscles all disjoined, their Cohesion perfectly ceasing."[25] Once the solid matter had dissolved into a common mass, it was "borne up in the Water" and was then precipitated back to the bottom in a new arrangement as the waters of the Flood receded. The laws of gravity determined the new positions of the fossils. "That for this reason the Shells of those Cockles, Escalops, Perewinkles, and the rest, which have a greater degree of Gravity, were enclosed and lodged in the Strata of Stone, Marble, and the heavier kinds of Terrestrial Matter: the lighter shells not sinking down till afterwards, and so falling amongst the lighter Matter, such as Chalk...."[26] Woodward's theory could explain the absence of fossilized land-based animals and plants in the strata and the re-growth of plants after the Deluge. Human bodies, trees, shrubs and other such matter are lighter than chalk

---

[25] Woodward, 74.

[26] Ibid., 76.

and the other rocks and minerals, so it precipitated last and laid atop all of the other layers. Consequently, they were "more nearly exposed to Air, Weather, and other Injuries: the Bodies of the Animals would suddenly corrupt and rot: the Bones, Teeth, and Shells, would likewise all rot in time, except those which were secured by the extraordinary Strength and Firmness of their Parts...."[27] By the same argument, the "Seeds of all kinds of Vegetables, being by this means reposed, and as it were, planted near the Surface of the Earth, in a convenient and natural Soil, amongst Matter proper for the Formation of Vegetables, would germinate, grow up, and replenish the face of the Earth...."[28] The Flood could explain, then, not only the placement of the fossils, but the chaotic structure of the earth's surface, with all of its mountains, valleys, ravines, and other "imperfections."

To Woodward, the only conceivable reason for the Deluge was that God had intended for a complete reformation of the Earth, and not merely an eradication of the corrupt humans living upon its surface. If the Deluge had been designed simply to destroy humankind, Woodward argued, it "might have been taken off at a far cheaper rate: without this ransacking of Nature, and turning all things topsie-turvy: without this battering of the Earth, and unhinging the whole frame of the Globe."[29] Wars, for example, would have been a very good way to exterminate the human race. Pestilence, or "a thousand other Disasters, whereby he could have carried them off by sholes, yea swept them all clear away."[30] Or he might have destroyed them by fire from heaven or holy thunder. All of these methods would have cost far less effort, and would have been quite effective in

---

[27] Ibid., 78.

[28] Ibid., 79.

[29] Ibid., 91.

[30] Ibid., 91.

their purpose. Yet they did not happen, so there must have been a greater reason for the Deluge, something beyond the purpose stated in the Bible. According to Woodward's vision, the universal Deluge was not "solely leveled against Mankind, but principally against the Earth that then was; with design to destroy and alter that Constitution of it, which was apparently calculated and contrived for a state of Innocence: to fashion it afresh, and give it a Constitution more nearly accommodated to the present Frailties of its Inhabitants."[31] The new conditions present in the postdiluvian Earth were reflected in the lives of the people that followed Noah. The life spans of modern men were now readjusted to the new form of the Earth to approximately one hundred and twenty years, mostly due to the soil's diminished fertility. There was only so much soil left near the surface "as might just sufficiently satisfie the Wants of human nature, but little or no more...but so that is should require Industry and Labour to excite it, and not yield a competent Crop without Tillage and Manure."[32] The seeming chaos and disorder of the modern Earth, then, were the result of the Deluge; "all the Irregularities and Inequalities of the Terrestrial Globe were caused by this means: date their Original from this Disruption, and are all entirely owing to it."[33]

An Essay toward a Natural History of the Earth was intended to show the absolute necessity of a God active in the Earth's creation and active in its history. Woodward's vision allowed God a great deal of direct action; God's power was not bound by natural law, and miraculous or providential events did not abide the limitations of natural law. Those critics who complained that Burnet took away too much from God with his natural theory of the Deluge were much more inclined to accept Woodward's view of the catastrophe. Woodward also leveled

---

[31] Ibid., 84.

[32] Ibid., 90.

[33] Ibid., 80.

his argument against the atomists and the threat of atheism. "Nor much more wonder is that an *Epicurus*: one who could ever espouse a Notion so enormously absurd and senseless, as that the World was framed by Chance: that this vast, regular, and most stupendous Pile was owing to no higher Principle than a fortuitous Congress of Atoms: and that either there was no God at all, or, which is much the same thing, that he was an impotent and lazy *Being*, and wholly without concern for the Affairs of the lower World."[34] According to Woodward, then, the random motion of atoms could not explain the order of the universe; it was the work of God.

Many naturalists found Woodward's effort to eradicate materialism and atheism admirable, but some, like Vallisneri, who had extensive experience with the strata, were critical of Woodward's theory.[35] As Vallisneri wrote to Bourguet in August of 1721, he believed that "Woodward has taken the words of the Holy Scripture and the most Sacred Fathers with too much rigor, and has not considered that they have spoken hyperbolically so as to frighten the populace, and he has almost forgotten to look at the book of nature that is never wrong and is less difficult to be a good judge of than the Holy Scripture."[36] According to Vallisneri, the book of nature proved Woodward wrong. Fossils were not arranged in nature by specific gravity, and the larger fossils did not necessarily reside in the lowest strata. Instead, they appeared in order of increasing complexity, as if the strata were laid down in chronological order of development. But even with all of its faults, Woodward's work gained wide acceptance because it integrated scripture and reason in a way that rendered his theory acceptable to the widest range of scholars. Perhaps its strongest argument was for the organic

---

[34] Ibid., 58.

[35] Elias Camerarius was another who questioned Woodward's thesis. See Oldroyd, 54.

[36] Vallisneri to Bourguet, 30 August 1721, BVN, Ms. 1282.

origin of fossils, and if one could not accept Woodward's theory about the Deluge, one could at least agree that the fossils were the remains of once-living sea animals.

The Flood's Purpose

Aside from the difference of opinion about the relative merits of the book of nature and the Holy Scriptures, Vallisneri also disagreed with Woodward about the purpose of the Flood. The main purpose of the Deluge, Vallisneri believed, was to eliminate sinful humanity (*mal nata gente*), "not to ruin the foundations of the Earth."[37] Like Leibniz, Vallisneri believed that the fundamental underlying structure of the Earth had remained immutable since its creation. Vallisneri could not bring himself to believe that the Deluge had destroyed the grandeur of the mountains and completely restructured the Earth's surface.

> Must we believe that it is an error, or a scourge, or an effect of the same, that the strata are not all horizontal, nor tilted the same way, yet concede that before their generation they could not have been located so? This variety of position is a most subtle artifice of the hand of the great master God – which seems a defect to our short vision, or a scourge, or accidental disgrace – is a most wise order to obtain various ends, which I am not going to describe, because it would be too long and too tedious.[38]

Our limited vision was insufficient, according to Vallisneri, to understand the entirety of the great Master's creation.

Vallisneri saw problems, too, in the natural mechanism of Woodward's argument. If, as Woodward argued, the entire surface of the globe became liquid, would not the fossils themselves break down and loose their cohesion? How could it be that the fossils survived when much more solid objects like mountains

---

[37] Vallisneri, *De' Corpi marini*, 49.

[38] Ibid., 66-7.

were destroyed? And "at what time during the fatal Deluge did this miracle [the layering of the fossils] happen? In the beginning, the middle, or the end?" It could not have been the beginning, Vallisneri argued, because the mountains were still uncovered and the waters would have simply lapped up against their foundations. It could not have happened in the middle because the Earth would have been reduced to chaos, a world without order or structure: a world of "avenging murky waters, tumultuous, [and] angry," sent by the Creator. And how, at the end of the Deluge with all of its terrible transformation, did the fossils that were on the mountains and hills return to their places?[39] In Vallisneri's, opinion, then, Woodward's mechanism for fossil placement was faulty in both conception and execution.

According to Vallisneri, there was also a shortage of water for the truly universal Flood envisioned by Woodward. Natural historians working in the late seventeenth century commonly accepted the idea that forty days of rain was not enough cover the globe in water. Burnet and Woodward both argued for a Great Abyss, a network of subterranean caverns, provided sufficient water, in addition to that of the rain, to cover the highest mountain peaks. In their favor are two passages from Genesis: "In the six hundredth year of Noah's life, in the second month, the seventeenth day of the month, the same day were all the fountains of the great deep broken up, and the windows of heaven were opened."[40] "And God remembered Noah, and every living thing, and all the cattle that was with him in the ark: and God made a wind to pass over the earth, and the waters asswaged; the fountains also of the deep and the windows of heaven were stopped, and the rain from heaven was restrained."[41] Vallisneri, in spite of these Biblical references,

---

[39] Ibid., 27-28.

[40] Genesis 7:11-7:12.

[41] Genesis 8:1-8:2.

doubted the existence of the Abyss. "Where did it open," he asked, "this horrendous chasm that absorbed half a world of water and gulped it down into the Earth? And why is the rift, or at least something of its vestiges, not visible [now]?"[42] He also found it difficult to believe that a huge mass of water could coexist with the hot magma found in the Earth's crust. Would not the water vaporize as a result of the heat? In Vallisneri's opinion, the only available source of water was the heavens, and this was only enough to provide a limited flood.

In addition to the inadequacy of the water supply, Vallisneri saw a problem in the capacity of the Ark. If "the Flood were truly universal, it would be necessary that Noah save in the Ark all the families and all the species of insects," because if they perished it would "break the most noble succession, and chain of all living beings."[43] In the sixteenth century, Athanasius Kircher had explained away the problem of the insects by arguing that they spontaneously generate from decaying matter (the Aristotelian position), so it was not necessary to have them on the Ark. But in the seventeenth century, Francesco Redi had shown that insects come from eggs just like other animals do, and Vallisneri insisted that there must then be consideration for them on the Ark. So the Ark must have held a small sample of every insect then known – a huge quantity to be sure. In combination with the animals, their weight would have put a severe strain on the Ark's ability to sail. Moreover, the animals and insects living there must have something to eat. Think about how much food the Ark must carry for all these insects, Vallisneri pondered, all the "grass, roots, fruit, flowers, and grain."[44] How would it be possible to conserve all of these plants? Some authors, like

---

[42] Vallisneri, *De' Corpi marini*, 91.

[43] Ibid., 85. For the significance of the "chain of being" in early modern thought, see Arthur O. Lovejoy, *The Great Chain of Being: A Study of the History of an Idea,* (New York: Harper, 1960).

[44] Ibid., 85.

Burnet, argued that the seeds could survive underwater, and they simply grew once the Flood receded. But in Vallisneri's opinion it is likely that the water would carry them away to soils in which they will not grow.[45] Aristotelians, on the other hand, in an argument similar to Kircher's for insects, argued that many plants did not need seeds; they would grow spontaneously out of the earth by themselves. In response, Vallisneri pointed out that he (in other work) and others had shown conclusively that all plants need seeds to grow. It was impossible that the Ark could have carried the entirety of the Earth's living beings for a year, including plants, and then repopulate the Earth after the Floodwaters had receded.

Holy Scripture and the Holy Laws of Nature

Vallisneri believed that the Flood could be explained by a "more natural and simpler cause, one more in conformity with the laws of the Great Mother who, even in her oddities and her errors, knows her own limits and neither knows how to nor is able to escape out of those limits, except on the explicit command of the Most High."[46] From nature, the most probable argument was that the Deluge was only "universal" in Asia. The population of the Earth had been concentrated in that area, and since the purpose of the Flood had been to destroy humanity, it was only necessary for the Flood to cover the inhabited zones. The rest of the Earth could be spared. A limited Flood would greatly reduce the number of plants, animals, and insects that Noah would have had to carry and would make the biblical narrative much more likely from a naturalist's perspective. Moreover, Vallisneri's observations showed that the fossilized marine animals were saltwater varieties, which effectively countered scholars who argued that the fossils were creations of the Flood itself, because in that case they would have been freshwater

---

[45] Ibid., 88.

[46] Ibid., 31. This translation from Rossi, 76.

varieties.[47] But according to Woodward's theory, fossils had been created before the Deluge and were merely *repositioned* by it. Yet if this were the case, Vallisneri argued, "marine materials should be found universally in the ground." In his "Mountain and Alpine travels," Vallisneri found marine fossils only below a certain altitude and on mountains facing the sea.[48] Moreover, as his observations of the strata confirmed, the fossils therein do not follow a pattern of heaviest (at the bottom) to lightest as Woodward's theory predicted. Vallisneri could not accept the diluvial theory at any level; it neither explained the origin of the fossils nor their placement.

In Vallisneri's view, the evidence indicated that the formation of fossils occurred over a long period of time. There were, he argued, "many particular inundations – that was the judgment of Plato and many Learned Greeks – and recessions of the Sea, and other causes less monstrous, more probable, more simple, and more natural [than the Deluge]. So have noted many erudite Moderns, who recognize the oldness of the World and the varied methods of change on its wrinkled exterior face."[49] Vallisneri believed – like Leibniz – that these inundations once covered the vast plain of the Po valley, with only the peaks of the mountains around Modena and Reggio uncovered.[50] As evidence he pointed to the structure of the strata and their composition. "Anyone who is not blind can see that the structure of the mountains, where they are worn down to the rock or laid bare, is very different from that of the plains, where deep wells are dug." First there is a layer of round stones, then one of smaller stones, followed

---

[47] Ibid., 19.

[48] Ibid. This observation Vallisneri had made much earlier, in a letter to Bourguet dated 9 November 1710, BVN, MS. 1282, fols 16-17.

[49] Ibid., 34.

[50] Ibid., 39.

by a sand layer, and finally a layer of earth. This order is repeated "again in other places, and again above these, in other places up to the top of the mountain, which shows clearly that the mountain was made in several different times and by several inundations, not just by one."[51]

The strata that were laid down by these many inundations were, for Vallisneri, a reflection of God's plan and the natural order of the universe. Vallisneri did not explicitly describe his beliefs about the Earth's origins in *De' Corpi marini*, so it is difficult to know exactly how he envisioned God's role in creation. It is clear, though, that whether the initial process of creation was due to natural forces or divine will or some combination of the two, he believed that the subsequent epochs in its history were governed by natural law. "God…formed the world in layers, and he also set the laws for the elements, for motion, and for nature, so that his fundamental and essential layers would always be conserved, and so that alterations, which we see occurring through subterranean fires and other causes, would all be directed to his purposes." This regenerating power guaranteed that "the first bases would ever be solid, and the primal, immeasurable framework of the mountains would never be unhinged or removed."[52] The world is somewhat like the human body, undergoing a constant process of renovation that preserves its basic form and function without the necessity of supernatural or divine intervention for its orderly function. The Earth's surface might change over time, "the immortal Galileo imagined such endless mutations in the surface of the Earth, that the inhabitants of the Moon would be able to see them," but it was a natural mechanism that retained nature's balance.[53] Vallisneri's vision of a world with a long history, in which natural processes like erosion and vulcanism

---

[51] Ibid., 56-57. Translation from Rossi, 77.

[52] Ibid., 73-74. Translation from Rossi, 77.

[53] Ibid., 44.

slowly shaped the Earth's surface over thousands of years, anticipates the theory of uniformitarianism promoted, almost a century later, by James Hutton and Charles Lyell.

Antediluvian Men

In Vallisneri's "Second Letter" there is a seeming digression, strange to the modern eye, on the lengthy life spans of antediluvian men, which, according to the Bible, often exceeded five hundred years. Why did Vallisneri include the topic in his book on fossils and the Flood? Was it a form of self-censure to avoid the ire of the Catholic Church? Considering the openness of Vallisneri's earlier reliance on natural law and arguments against the universal Deluge, it seems odd that he would seek the shelter of orthodoxy on this topic, and although it may seem counterintuitive, I would like to suggest that Vallisneri's acceptance of 900-year antediluvian life spans actually was an *unorthodox* position, one that undercut the more accepted view of the Earth's history.

Vallisneri's contemporaries who wrote about history generally treated antediluvian historical claims skeptically, mostly because their view of a short terrestrial history (about 6,000 years) and the youthful nature of human civilization rendered a lengthy antediluvian epoch impossible. If, as they believed, the world was created four thousand years before Christ, and if the Deluge happened roughly 1,400 years afterwards, then there was no time to allow successive generations of antediluvian men with life spans in excess of five hundred years. For the majority of scholars who accepted the short chronology of the Earth's history, like Benoît de Maillet, even biblical claims about the pre-Noachian period were considered fabulous.[54] De Maillet went so far as to claim the patriarchs actually had shorter life spans than modern humans, because they

---

[54] Rossi, 167.

lived in caves and, lacking agriculture, had to rely on wild herbs and fruits for their sustenance.[55]

Vallisneri, considering his views on the Deluge and antediluvian natural inundations and slow retreat of the waters, seems clearly in opposition to this "short history" point of view. In a letter to Muratori written in October of 1721, he wrote, "if you read with attention the first and second dissertations [First and Second Letters] you will observe that the explanation of marine bodies that are found on the mountains and the system about the creation of the world in which we must have faith [the Mosaic account], are full of unsolvable knots."[56] In *De' Corpi marini*, Vallisneri described the Earth as "very old," and his conception of the Earth's past, with its slow process of natural actions such as volcanism and erosion, suggested a much longer history than imagined by theologians and naturalists who supported the biblical account of history. In this context, Vallisneri's acceptance of the long lives of the biblical patriarchs seems less paradoxical. If history extended tens of thousands of years into the past, then nine-hundred-year life spans were easily accommodated. Vallisneri was part of a small, but growing, group of natural historians who believed that the Earth's history was much longer than the traditional six thousand years and that it has been shaped by natural causes. In the eighteenth century this was a minority opinion, but the evidence for it was mounting, and by the nineteenth century the majority of naturalists would accept it.[57]

---

[55] Benoît de Maillet, *Telliamed*, vol. 2 (Amsterdam, 1748), 61-62.

[56] Vallisneri to Muratori, 8 October 1721, in L.A. Muratori, *Edizione nazionale del carteggio di L.A. Muratori*, vol. XLIV, edited by Michela L. Nichetti Spanio (Florence: Olschki, 1978) p. 179. See also, Tucci, Francesco Saverio. "Il parlare della S. Scrittura e l'operare della natura: gli interrogativi della geologica storica nella riflessione di Antonio Vallisneri." *Contributi* 7 (1983), 27.

[57] Benoit de Maillet, in his *Telliamed* of 1748, argued that the earth's age could be billions of years. Most natural philosophers strongly opposed this view. Georges Luis Leclerc, Comte de Buffon, in his *Epoques de la nature* of 1778, estimated the earth's age at seventy-five thousand years. His estimates were highly controversial. By the nineteenth century, though,

If the biblical account of the antediluvians was accepted as fact, or at least as possible, then why did they live so long? "Some believe," Vallisneri wrote, "that the long lives of the first Fathers and Patriarchs was an act of Divine Providence, to quickly fill up the World with men, ...but this is a moral reason, not physical, as you [Father Leoni, recipient of the "Second Letter"] have researched."[58] Vallisneri, of course, would not be satisfied with moral reasons; he had to find the physical causes to truly understand the phenomenon. As had been his approach in the "First Letter," Vallisneri initially addressed the arguments of other scholars so that he might critique them. Thomas Burnet's vision of the antediluvian era, Vallisneri argued, with its sweet temperate air and soil that would produce fruit without "bathing one's brow in sweat," was poetic, but hardly realistic.[59] Theories such as these were jokes, he wrote, "to tell naïve little boys and sweet simpletons, not those who know the true History of Nature, and, indeed, the most strict justice of the Lord that requires Adam and his Descendants to pay the penalty of disobedience, from which the Antediluvians are not exempted."[60] Some other scholars adopted a more pragmatic solution to the problem, arguing that the years being counted in the Bible were lunar and not solar. In response, Vallisneri called upon the authority of Lioni, Burnet, and Don Giovanni Orsato, a "great ornament of the Benedictine Religion, grand Theologian, and great Scholar, who was meritoriously granted by our most prudent Riformatori of the Studium the *Chair of Sacred Scripture*."[61] All three had verified that the years counted in the bible were indeed solar, not lunar, and

natural philosophers received quite favorably Charles Lyell's argument, in Principles of Geology (three volumes, 1830-1833), that the earth was at least several hundred thousand years old. See Oldroyd, 134-135 and Rappaport, 250-262.

[58] Vallisneri, *De' Corpi marini*, 112.

[59] Ibid., 101.

[60] Ibid., 101.

[61] Ibid., 105.

Vallisneri was satisfied with the quality of their proofs. "If, then, the years were Solar, if there were the rotations of the seasons, if the fabric of our body was then as now, and now as then, if woman must give birth with pain and many dangers, if it was necessary that man pour out the same hard work for sustenance as the disobedient Adam, and his descendants, by command of the Most High, if the antediluvians were like Us, and perhaps more than us in every manner of vice," Vallisneri wrote, it would take much reasoning to discover the cause of their longevity.[62] The cause, moreover, must conform to the laws of medicine and nature.

Vallisneri's solution was not to explain why the antediluvians lived so long, but to explain why we do not. His explanation hinged on the poor quality of air, food, and drink caused by environmental factors in the post-diluvial era, and poor heredity in modern humans caused by psychological factors. The environmental factors were the result of the atmosphere becoming, "corrupt, very heterogeneous and [filled with] tumultuous particles" from the Flood.[63] Poor heredity was also the result of the Deluge, caused by

> the terror of people seeing themselves in the midst of so much water, the imprisonment, so to speak, for a year, the terrible [fact of] the life of a whole world submerged, the anguish of [seeing] kin killed, that natural repulsion and suffering of the soul at [the sight of] a spectacle so deadly, so dismal, so horrifying, so disgusting [as] innumerable corpses of both men and cattle; the horror, even after the desolated desert – [all this] without any doubt contributed much to destroying the perfect harmony of the blood and disturbing to the regular movements of their spirits and their humors. Thus the very principles of generation were impaired, and the defects penetrated them so very deeply that they were passed on to their unborn children, and the hereditary misfortune… will last to the end of all time, because, though it is now enormously

---

[62] Ibid., 106.

[63] Ibid., 107.

> multiplied and divided throughout the world, we all come from that infected lineage.[64]

Vallisneri believed that the fruit, grass, grain, and air all returned to their previous states once the post-diluvial humidity had cleared. In this way the Earth was restored to it antediluvian form. It was too late to restore humankind, however, and the hereditary effects of the Flood were permanent. The Deluge, then, in Vallisneri's vision, was not universally catastrophic from a geological standpoint, but humanity never recovered from its effects.

Libertinism and Self-Censorship

Although Vallisneri's examination of the long life spans of the antediluvians was consistent with his views about the long history of the globe, the "Second Letter" still presents a puzzle because it contains a mixture of naturalistic and supernatural explanations for geo-historical events. The problem is clearly illustrated in the contrast between the way the letter begins and the way it ends. In the first paragraphs of the "Second Letter," Vallisneri professed his belief in the universality of the Deluge. "I establish, then, the truth of the Universal Deluge, as authenticated by Holy Writ, because it would be too unbecoming to contradict, or explain differently, a thing so beneficial to our souls."[65] In the time of the Deluge, he went on to write, there were many miracles, and miracles had properties completely outside the laws of nature. Thus, by divine and miraculous means, all of the plants necessary for the renewal of the post-diluvial Earth could be conserved, and indeed multiplied. Yet Vallisneri ended the "Second Letter" with a prediction that natural philosophy might, in the future, result in "a praiseworthy System of truth that explains every

---

[64] Ibid., 108. The translation is from Rossi, 79.

[65] Ibid., 82.

occult phenomenon of Nature." How could Vallisneri on one hand advocate a natural philosophy that will "explain every occult phenomenon of nature," and on the other hand accept a miraculous Flood exempt from natural law? This contradiction was puzzling even to Bourguet, Vallisneri's friend and long-time correspondent, who in 1722 complained that he remained unsure about whether Vallisneri believed the Flood was miraculous or natural.[66]

One common historiographical approach is to argue that Vallisneri did not truly believe in the miraculous nature of the Deluge and had self-censured the work to avoid suppression by the Catholic authorities.[67] According to this argument, Vallisneri's acceptance of the long lives of the antediluvians, with its explanation of the phenomenon in natural terms, was a stratagem that allowed him to lay bare his true convictions without raising the ire of the censors. It is focused on Vallisneri's natural explanations and assumes that he did not truly believe that the biblical narrative was literally correct on all points or that naturalists should use it uncritically. A look into Vallisneri's correspondence, where he could more openly state controversial opinions, reveals some evidence to support this argument. In November of 1717 Vallisneri wrote to Louis Bourguet: "Except for the faith that one must have in the Holy Scripture (which is used for every dispute, and we are talking as natural philosophers), who is assured of this Universal Deluge?"[68] Three years later he reiterated his anti-diluvial position, "you and many other learned men hold firm [on the Universal Deluge], because the Holy Scripture says it's so... but the Holy Scripture teaches natural philosophers

---

[66] Bourguet to Abauzit, 28 February 1722, BVN, MS. 1259. This letter is cited in Rhoda Rappaport, "Italy and Europe: the case of Antonio Vallisneri (1661-1730)," *History of Science* 29 (1991) 87.

[67] See Rappaport, "Italy and Europe," 86; Dario Generali, "Fra Filologia e storiografia: Progetti di edizioni settecentesche," *Studi Filosofici* 18 (1995) 89-90; and Tucci, 9.

[68] Vallisneri to Bourguet, 23 November 1717, BVN, Fonds Bourguet (ref. 8), MS. 1282, fols 226. Tucci says this letter is actually from 23 November 1718.

nothing; it teaches the ways of Heaven, not the phenomena of the Earth."[69] There is also Vallisneri's letter to Bourguet from 1 January 1722 wherein he writes, when "we resort to miracles, everything is provided to Natural History. I have used them truly often in my work. But do you know why? Because the priests are silenced, otherwise I intend to say that those things did not happen in the way imagined by Woodward and the others with him."[70] Further evidence can be found in the letter Vallisneri wrote to Bourguet on 30 August 1721, in which he highlighted the Galilean theme of double revelation, biblical and natural. "Nature is infallible and is the voice of God," he wrote, "with this difference, that the language of Holy Scripture can and should be interpreted in many ways (otherwise it would say many things contrary to the evidence of the senses), but the language of Nature is always the same, without metaphor, without allegory, without hyperbole, without doubtful, obscure, mysterious meanings. Nature speaks clearly to him who knows how to understand her, and has no need of interpretation."[71] In the same letter, he mentioned the argument about the antediluvians, "of the years of the Antediluvians, I introduce, with art, the strongest reasons against the answers that they are able to give." These letters do seem to indicate that Vallisneri was hiding behind a kind of intellectual smoke screen, so that the censors would approve his work without concern for the overly naturalistic explanations contained therein.

In eighteenth-century terms, this would make Vallisneri a libertine, a word that was used to describe those who rejected Christian revelation as the basis for

---

[69] Vallisneri to Bourguet, 23 September 1720, BVN, Fonds Bourguet (ref. 8), MS. 1282, fols 237-8.

[70] Vallisneri to Bourguet, 1 January 1722, BVN, Ms. 1282, fols 251. Also see note 350, in Giovanni Artico di Porcia, *Notizie della vita, e degli studi del Kavalier Antonio Vallisneri*, ed. Dario Generali (Bologna: Patron, 1986), 137.

[71] Vallisneri to Bourguet, 30 August 1721, BVN, Ms. 1282, fols 249-250. This translation appears in Rappaport, "Italy and Europe," 84. Generali's reference to it can be found on page 137 of Di Porcia's *Notizie della Vita*, in footnote 350.

morality and politics and instead looked to reason and nature for the foundations of society.[72] Vincenzo Ferrone, who is perhaps the most polemical of those who have described Vallisneri's views on nature and religion, has gone so far as to categorize Vallisneri as an "intellectual militant in favor of modern, secular, Galilean science based on his hypothesis of mechanism."[73] Libertines were skeptical of divine revelation, applied critical thinking to philosophical and spiritual pursuits, and were aware of the historical nature of religions and their mythologies. Giordano Bruno and Tommaso Campanella are good examples of Italian libertines; Bruno, who was burned at the stake in 1600 for heresy, agreed with Copernicus on the structure of the solar system and believed in an infinite universe of inhabited worlds, and Campanella, a contemporary of Galileo, argued that Holy Scripture did not reveal nature's secrets. Vallisneri's irreverent attitude about biblical exegesis and biblical authority on matters of natural history seem to support this portrayal of his libertinism. Due to their skeptical approach to religion, libertines were generally either atheistic or deistic, and so to include Vallisneri in their numbers the historian must assume that Vallisneri's pious statements were not genuine. On the other hand, if Vallisneri's "protestations" of piety were authentic, then it is problematic to categorize him as a libertine.[74] Certainly Vallisneri was dedicated to the "holy laws of nature," but he also admired the work of God's creation, and the divine wisdom that is reflected in

---

[72] Due to its eighteenth-century use as a label for just about anyone with unconventional ideas, the term is extremely difficult to define. See Rappaport, *When Geologists Were Historians*, 212, for the example of Fontenelle.

[73] Vincenzo Ferrone, *The Intellectual Roots of the Italian Enlightenment*, trans. Sue Brotherton (Atlantic Highlands: Humanities Press, 1995), 106.

[74] See Rossi, 75-79. Giuseppe Montalenti, in his entry on Vallisneri in the *Dictionary of Scientific Biography*, also accepted Vallisneri's piety. Giuseppe Montalenti, "Antonio Vallisneri," *Dictionary of Scientific Biography* 13 (1976): 562-5.

physical reality.[75] A reading of the book of nature, then, was also a reading of the holy word, and in this context the "holy laws of nature" was not a paradoxical phrase.

Ultimately, though, the question of Vallisneri's "true" opinions reveals more about twenty-first-century historians than it does about early modern science. It is true that naturalists like Vallisneri worked in a culture in which they had to be extremely careful about their opinions, because they might have been interpreted as heretical or atheistic. Scholars more in line with the Christian tradition and the Catholic Enlightenment roundly criticized any theory that had mechanist overtones. But Vallisneri and Galileo considered themselves good Catholics, and to portray them as heroes of secular science is problematic. The separation of science and Christianity did begin to some degree with Galileo, one must not underestimate the impact of his legacy, but natural history in the early eighteenth century was far from atheistic. Physico-theology, or natural theology, was an important current in natural history until the middle of the nineteenth century.[76] Trying to understand Vallisneri's "true" purpose, and designating it as secular, modern, and anti-Catholic, in accordance with our desire to see him as a worthy precursor of our modern values, does little to enrich our understanding of his culture. Moreover, it implies that he was a kind of progressive reformer in the mold of the Enlightenment *philosophes*. There is no evidence to show that Vallisneri was attempting to secularize society; it shows only that he held that Holy Scripture should not have precedence in the worldly realm of natural history. Considering the secular nature of modern science, this may have had a great long-term significance, but it should not be taken for an advocacy of a larger scale

---

[75] Rossi, 78. Rossi takes his information from the second edition of *L'Origine delle Fontane*. Antonio Vallisneri, *Lezione academia intorno l'origine delle fontane*, 2nd ed. (Venice: P. Poletti or A. Bortoli, 1726), 7-8.

[76] Rappaport, "Italy and Europe," 88.

Enlightenment style libertine reform. Since Vallisneri was primarily concerned with the worldly, rather than the religious, it is not surprising that he was more focused on the natural aspects of the Earth's past than the supernatural ones. But his belief that the scriptures should not be applied to natural phenomena does not imply a categorical rejection of all Catholic teaching. Vallisneri's spiritual beliefs mirrored Galileo's opinion that the Church should hold authority in matters of faith, and that science should hold authority in matters of nature. In any case, Vallisneri succeeded in his approach. Unlike his book on human generation, published the same year, which was held up by the censor, *De' Corpi marini* passed through the censor's hands without eliciting a negative response.[77] It is probably best to conclude, then, that Vallisneri was a Catholic naturalist in the Galilean mold who believed in the "holy laws of nature" and who was a realist enough to work within the cultural constraints of early eighteenth-century Italy.

Conclusion

Vallisneri's study of the fossils, then, touched on the fundamentals of theology and philosophy. Fossils were suggestive in that they were remnants of a long distant past, perhaps left from the days of the creation or from the cataclysm of the Deluge, which was most often portrayed as a divine, supernatural event. Since the stakes were so high – the natural socio-political order and the foundations of Christianity were potentially under critical scrutiny – any unconventional theory on their origin could be seen as a threat to hallowed traditions of Christianity. Some seventeenth-century natural philosophers, like Newton, believed the safest route to take was to declare that the act of creation itself was a miracle, and thus exempt from natural law. Others, like Thomas

---

[77] The book on generation is Antonio Vallisneri, *Istoria della Generazione dell'Uomo e degli Animali se sia da Vermicelli spermatici o dale Uova* (Venice: G. Hertz, 1721). The censor held up the book because he disagreed with Vallisneri's ovism. Vallisneri's powerful patrons quickly came to his aid, and the censor relented.

Burnet and John Woodward, attempted to reconcile the Mosaic account of creation with the natural evidence to create a kind of sacred history or physico-theology that would protect the Divine Word. This was a dangerous endeavor because it allowed reinterpretation of the Bible based on natural findings, and it took the some authority for biblical exegesis away from theologians and placed it in the hands of naturalists.

Vallisneri mirrored Burnet and Woodward in that he, too, used natural forces in his explanations for biblical events, but he rejected its universality and its role in fossil placement. In Vallisneri's opinion, the Flood was local to Asia, the part of the globe inhabited in the antediluvian era, because it had not been necessary for it to be truly global. God intended it to wipe out humanity, not to reform the Earth. One of the arguments Vallisneri used to support his criticism of Woodward's diluvianism was that Noah's Ark could not have carried all of the animals necessary to repopulate the earth after the floodwaters receded. Vallisneri suggested that all the varieties of insects, for example, which were an integral part of the great chain of being, would have filled the Ark, leaving no room for the other links in the chain. He did not propose an alternative to Burnet's or Woodward's "world system;" Vallisneri's focus rested specifically on the issue of fossil formation, which he believed was the result of natural inundations over the course of a very long time span. The fossils were deposited in an epoch when the mountains were covered by water, which in succeeding ages receded, leaving them far from the sea. His evidence was the fossils themselves and their placement within the strata, which clearly contained alluvial rock; there was no need for divine intervention in any of these natural processes.

Considering the naturalistic and skeptical approach he took to the Flood, one wonders if he chose to address the long lives of the antediluvians in the "Second Letter" of *De' Corpi marini* in order cover up his libertinism and avoid censorship. Although there is evidence in his correspondence that suggests Vallisneri resented the necessity of pandering to the Catholic censors, it does not appear that he was a true libertine. Although there is a confusing mixture of natural and supernatural explanations in the "Second Letter," Vallisneri's final conclusions about the antediluvians were remarkably consistent with the rest of his natural philosophy. To accommodate the incredibly long lives of the antediluvians, a longer time between creation and the Flood would be necessary than was allowed in the traditional timeline of 1,400 years. The book of nature, in this case, supported the holy book. Although his announced belief in a miraculous universal Flood may have been somewhat insincere, there is no indication that Vallisneri disagreed with the Catholic Church on issues of faith or that he felt there was a necessary incompatibility between natural philosophy and theology. The "holy laws of nature," as he called them, dictated the phenomena of the earth, and the Holy Scriptures the ways of heaven. Much like Galileo, then, Vallisneri was a good Catholic who believed that the Church should focus on matters of faith and leave worldly matters to the natural philosophers.

Michael D. Cunningham (University of Connecticut) earned his Ph.D. in history from the University of Connecticut in 2005. His dissertation examined the work of Antonio Vallisneri, an early eighteenth-century Italian natural philosopher. He has received grants from the Fulbright Foundation and the Gladys Krieble Delmas Foundation. He is currently an instructor at the University of Connecticut.

Loathsome Beasts: Images of Reptiles and Amphibians in Art and Science

Kay Etheridge

The mythology and symbolism historically associated with reptiles and amphibians is unequaled by that of any other taxonomic group of animals.[1] Even today, these creatures serve as icons - often indicating magic or evil - in a variety of media. Reptiles and amphibians also differ from other vertebrates (i.e. fish, mammals and birds) in that most have never been valued in Europe as food or for sport. Aside from some limited medicinal uses and the medical concerns related to venomous species, there was little utilitarian value in studying the natural history of reptiles and amphibians. Because of this history and other characteristics of these animals, the images of reptiles and amphibians played a unique role in the study of natural history from the Medieval through the Early Modern periods. The images I will discuss come from books that have been analyzed by other scholars, but in most cases there has been little or no scrutiny of the portrayal of the herpetofauna. Because much of my research as a biologist has focused on reptiles and amphibians, I will consider their differences from mammals and birds.[2] In doing so, I will address image content from a somewhat

[1] One of the first stories in Christian writings evokes a snake up to no good, and a plague of frogs was evoked as a curse in a later bible story. However, the use of these animals in mythology predates major religions by at least two millennia; an example is the constellation Hydra (a great serpent), recorded on a tablet as early as 2400 BC. Daniel Bell, "Herping the Night Sky: The Mythology of Snake and Lizard Constellations," *Sonoran Herpetologist* 18:6 (2005), 63-64. Some contemporary emblems also originate from myths involving reptiles or amphibians; for example, the rod entwined by snakes is used as an icon by the medical profession. The symbolic association of serpents with curative properties dates to Greek mythology. I. A. Ramoutsaki, S. Haniotakis, and A. M. Tsatsakis, "The Snake as the Symbol of Medicine, Toxicology and Toxinology," *Veterinary and Human Toxicology* 42:5 (2000), 306.

[2] Fish are more similar, as they also are "cold-blooded" animals and closer phylogenetically to herpetofauna than to the furred or feathered "warm-blooded" animals. For

different point of view than that of an art or science historian. My contention is that understanding the evolving portrayal of these "loathsome beasts" is particularly useful in tracing the development of the study of natural history. I also will address how changes in these images over time reflect a transformation in how nature was viewed and valued in western European culture.

Prior to the advent of photography, drawings and paintings of plants and animals were the primary means used to convey the appearance of an organism, short of a live or preserved specimen. In fact, images have some advantages over actual specimens. First, they can be replicated and disseminated with relative ease. In the case of animals that do not hold their form or color when preserved, the image is actually superior to the specimen in representing the organism as it appeared in life. The skins of amphibians (frogs and salamanders) and reptiles (snakes, lizards, turtles, amphisbeanids, and crocodilians) are particularly prone to losing color in preservatives like alcohol. In a sense, the image itself preserves the organism, and in the case of species now extinct, this form of preservation is particularly valuable to students of nature. Additionally, images hold other advantages over specimens; for example, they can depict life cycles, place the organism in its habitat, portray behaviors, and illustrate magnified details of structure.

The earliest natural history studies to be illustrated were manuscripts now known as herbals and bestiaries. Most of the known bestiaries were produced in England beginning in the twelth century, and they typically contained a variety of animals that would be well-known in Europe, exotics (e.g. lions, elephants, and crocodiles), and mythical beasts such as the basilisk.[3] Various "serpents"

---

example, fish, amphibians, and reptiles all have a very low metabolic rate relative to that of mammals and birds, and hence can survive much longer without food.

[3] The earliest illustrated bestiaries dating from the ninth century draw much of their information on both real and mythical animals from the Greek *Physiologus,* adding quotations from scripture and moralizations based upon the character of the animals. The natural history of the animals takes on a more predominant role in the text by the fourteenth and fifteenth century.

(serpentes) usually were depicted, including some with wings, and the turtle *(testudo)* and frog (*rana*) also figured in some bestiaries. The salamander (*salamandra*) was often depicted as a serpent-like animal as in the Aberdeen Bestiary (Figure 1),[4] although in some bestiaries the salamander does have four legs. The accompanying text in the Aberdeen Bestiary recounts the myth that salamanders can poison both fruit and water (hence the recumbent figure of a man sickened by the poison) and that they can survive fire and put it out. One suggested explanation for the origin of the fire myth is that salamanders often hibernate in logs, so that when wood is brought in and put on the fire, the animals suddenly appear, scuttling out of the log. *Salamandra salamandra*, the European or "fire" salamander discussed below can have bright yellow or orange patterns, perhaps reinforcing the connection with flames. Additionally, these salamanders do secrete a neurotoxin and can even spray this poison from pores along the back. Likewise some of the symbolism associated with snakes in bestiaries bears a relationship to aspects of their natural history. In many bestiaries a snake was pictured shedding its skin as it crawls through a crack in a rock or other narrow aperture. The image may show a characteristic snake behavior, but in the bestiaries this action was cast as an allegory of rejuvenation or salvation through shedding of the outer self.[5]

Bestiaries often served as pattern books for subsequent bestiaries, as well as models for the popular emblem books of the sixteenth century, and other works such as tapestries and carvings. Thus, these images were widely disseminated; copying was rampant and inaccuracies such as external ears on snakes and salamanders were perpetuated until some artists began to draw these animals from

---

Wilma B. George and Brunsden Yapp, *The Naming of the Beasts: Natural History in the Medieval Bestiary* (London: Duckworth, 1991), 1-9.

[4] The very well-preserved Aberdeen Bestiary (Aberdeen University Library MS 24) is believed to have been produced in the north of England around 1200.

[5] Ibid., folio 71r.

life. For reptiles and amphibians the trend toward more realistic representation appears to have been later than for birds and mammals, some of which appear as fairly accurate renditions as early as the sixth century.[6] Until the advent of printing, however, few people had the opportunity to view these images. The Renaissance brought not only printing but the artistic use of shading and perspective as tools for rendering a naturalistic appearance. Perhaps the most pivotal influence on Renaissance natural history images however was the new-found impetus to work "from life." Direct observation is particularly important when the item to be depicted is not well known; a skilled Renaissance artist could paint or draw the human animal without a model, but would have much more difficulty in producing a realistic image of a crocodile. Renaissance philosophers also moved toward observation and away from an exclusive dependence on classic texts for knowledge. In any case, exploration and expanding trade routes resulted in an influx of new and exotic creatures for which Aristotle, Pliny, and the bestiaries provided no context or imagery.

One of the earliest examples of an illustrated book that resulted from this confluence of circumstances was the travel account published by Pierre Belon (French, 1517-1564).[7] The quality of Belon's drawing can be seen in the woodcut of the chameleon (Figure 2). The chameleon is one of six reptiles illustrated in Belon's account, which includes four snakes and a Nile crocodile. Belon also portrayed a "winged serpent." The accompanying text states that the *serpents aellés* he observed in Egypt were embalmed specimens. Belon's image of one of these serpents, which "has feet" and is said to "fly from Arabia to

---

[6] The *Juliana Anica Codex* in Vienna is dated to 512 AD and contains detailed and identifiable images of a number of birds. Minta Collins, *Medieval Herbals: The Illustrative Traditions* (Toronto: University of Toronto Press, 2000), 39.

[7] Pierre Belon, *Les Observations De Plusieurs Singularités Et Choses Mémorables, Trouvées En Grèce, Turquie, Judée, Egypte, Arabie & Autres Pays Étranges* (Paris: Guillaume Cauellat, 1554). This book was first published in 1553 and was issued in several editions in the sixteenth century. Belon made clear that his writings were based upon his own "ocular observations," although like a good Frenchman he also included gustatory observations, and for at least one fish, a recipe.

Egypt," is not very detailed.[8] In contrast, although the image of the chameleon is not entirely accurate in every aspect, it is clearly identifiable to the level of genus in current taxonomy, and based upon the fact that he saw it in *Arabie*, possibly to the species, *Chamaeleo calyptratus.*[9] While this level of specificity in the image may be of interest to the modern biologist, it is Belon's careful observation of the living animal's grip on the branch that separates it from depictions of a possum-like ground dwelling "chameleon" that precede it in emblem books such as those by Andrea Alciato.[10] Belon's rendition shows precisely how this very large arboreal animal can cling so securely to its perch; chameleons have five toes on each foot that function in opposition to each other in a distinctive mechanism. The front feet have two toes that grip the outside of the branch and three on the inside, and the back feet are arranged in opposite fashion. Belon must have studied his subject carefully, and he clearly thought it important to represent this unique arrangement of digits accurately even though he does not discuss it in the text. Evidently others must have thought this a remarkable image, for versions of it appear in animal encyclopedias and even in subsequent books on emblematic animals, such as later editions of Alciato.[11] Ironically, Belon's image of the less-than-truthful winged dragon also appears in many of these books.

Belon's near contemporary, Conrad Gesner (Swiss, 1516-1565), largely overshadowed Belon's original contribution to natural history with his own prolific output in both botany and zoology. Gesner's five-volume *Historia animalium* was the first animal encyclopedia with figures drawn "from life" and the inclusion of these images is generally seen as a major turning point in animal

---

[8] Ibid., 133-134. Belon writes that he saw "*des corps embaumez tous entier de certains serpens aellés, qui ont pieds qu'on dit voller de la partie d'Arabie en Egypte,*" and that from this he made the "portrait."

[9] Ibid., 125.

[10] William B. Ashworth Jr., "Marcus Gheeraerts and the Aesopic Connection in Seventeenth-Century Scientific Illustration," *Art Journal* 44:2 (1984), 135.

[11] Ibid.

studies.[12] He produced many of these drawings himself, whereas others were copied from a variety of existing sources,[13] or were sent to him by contemporaries interested in natural history. Gesner's woodcut images are surrounded by extensive descriptive text that delineates the animal's natural history from Aristotle and Pliny as well as the associated legends, symbolic meaning, and the beast's name in various languages. However, Gesner was a keen naturalist and often added his own observations as well as textual information from contemporary contributors.[14]

The "loathsome beasts" were given much more attention by Gesner than by any of his predecessors. The second volume of *Historia animalium* is devoted to what Gesner termed the *quadrupedibus ouiparis* (egg-laying quadrapeds); these include amphibians and reptiles, although the snakes (many of which do lay eggs) were treated in his volume on serpents. Volume II contains images and text on turtles, a crocodile (copied from Belon), and various lizards. Some of these animals were drawn with a high degree of realism and can be identified to the level of taxonomic family or genus in current classification. One lizard was drawn in such detail that the individual scales can be counted, and the animal is shown both from dorsal and ventral views, a pictorial device used in modern zoology texts. Another image shows a precisely rendered sea turtle with the

---

[12] Gesner's volumes were published over a period of years; images and references herein are from volume two on the egg-laying quadrapeds (Konrad Gesner, *Historia Animalium Liber II: De Quadrupedibus Ouiparis* (Frankfurt: Roberti Cambieri, 1586)) and volume five on serpents, (Konrad Gesner, *Historia Animalium. Liber V: Qui Est De Serpentium Natura* (Zurich: Froschouiana 1587)). His place in the history of zoology has been noted by many scholars, see for example E.W. Gudger, "The Five Great Naturalists of the Sixteenth Century: Belon, Rondelet, Salviani, Gesner and Aldrovandi: A Chapter in the History of Ichthyology," *Isis* 22:1 (1934), 32-33 and Frank N. Egerton, "A History of the Ecological Sciences, Part 11: Emergence of Vertebrate Zoology During the 1500s," *Ecological Society of America Bulletin* 84 (2003), 207-208.

[13] For example, Gesner's chameleon is an almost exact mirror image of Belon's down to the position of the knot of the tree; Gesner does credit Belon for information on many of the animals in his encyclopedia. Gesner, *Historia Animalium Liber II*, 3.

[14] Volume four on the fish and other aquatic animals was particularly dependent upon a variety of sixteenth century naturalists for information and many of its 900 images. Unlike some encyclopedists who followed him, Gesner did credit his sources. Egerton, 207-09.

correct number of plates on its upper shell and a smattering of attached barnacles; this is clearly identifiable as *Eretmochelys imbricate,* the hawksbill sea turtle. This turtle was drawn from a dorsal view in order to depict the structure of the flippers and shell, but there is also a lateral view of the head showing the distinctive shape of the turtle's beak, a characteristic that separates it from all other sea turtles.

The amphibian images were equally innovative. Next to a mature frog that he termed *rana perfecta,* Gesner showed an incompletely metamorphosed froglet that had not yet lost its tail (*foetus rana caudatus*);[15] this image may have been the earliest visual description of amphibian development to be published. He also included images of four distinct types of frogs (given different names) and differentiated between aquatic and terrestrial anurans. Relative to the amorphous frogs depicted in the earlier bestiaries and emblem books, Gesner's images are astoundingly specific and information-laden. Most importantly, the move towards naturalistic representation of these animals demonstrates that curiosity about their structure was increasingly important to Renaissance philosophers, and that an emblematic image was not considered sufficient on its own. Gesner went further by providing both a very exacting image of a common European salamander and a contrasting drawing of a mythical salamander (Figure 3).[16] The former was clearly drawn from a live or fresh specimen of the previously mentioned *Salamandra salamandra* (this wide-ranging species occurs in Switzerland) and was rendered in great detail, giving it weight as an actual, observed animal. Gesner's salamander is equipped with the dorsal rows of pores

---

[15]Gesner, *Historia Animalium Liber II,* 81.

[16] Ibid., 81. Gesner's extensive accompanying text on the salamander is beyond the scope of this essay, but it is interesting to note that he reports his own experimental evidence that salamanders do not indeed survive fire.

that release poisons, just like the very similar image of the same species by Von Rosenhof, drawn over two-hundred years later (Figure 10).[17]

Gesner's fifth volume of *Historia Animalium* contains the serpents, both real and imagined. His images of snakes are strangely undefined and incorrect when compared to those of the frogs, lizards and turtles, and it may be that some of the snake woodcuts were based on drawings not by his hand.[18] Both volumes were published posthumously, and the sources of many illustrations are not documented, so the differences in image quality remain a mystery. Another puzzle is the inclusion of so many mythical creatures by a man who prided himself on working "from life." Much has been written elsewhere about the hydras, dragons, and basilisks of Gesner and subsequent encyclopedists, and I will comment on just two such here. Gesner included the portrait of an undulating mythical basilisk sporting a crown, a human-like expression, and a tongue shaped like an arrow, but as with the salamander, he also depicted a realistic snake near the image of the mythical one.[19] The latter is shown lying on the ground in a naturalistic pose, and although the head suggests a slightly crown-like shape, it still looks like a snake. Gesner also included a copy of Belon's winged serpent, and he quoted Belon's text on it and cited him.[20]

The attention given to these images of reptiles and amphibians indicates that Gesner was interested in their truthful portrayal as much as that of the more familiar mammals and birds in volumes one and three of his encyclopedia. This

---

[17] August Johann Rösel von Rosenhof, *Historia Naturalis Ranarum Nostratium: In Qua Omnes Earum Proprietates, Praesertim Quae Ad Generationem Ipsarum Pertinent, Fusius Enarrantur* (Nuremberg: Johann Joseph Fleischmann, 1758), frontispiece. The von Rosenhof salamander image was made from a copper-plate engraving and therefore can include more fine detail than a woodcut. The eighteenth century salamander image is complete with the characteristic pores on the parotid gland behind the eye, but other than this and the fact that it was hand-colored, it offers little information not found in Gesner's fine woodcut.

[18] Gesner, *Historia Animalium. Liber II, passim.*

[19] Gesner, *Historia Animalium. Liber V*, 32.

[20] Ibid., 55.

emphasis on was realism is a significant change from the countless inaccuracies and lack of specificity in images of reptiles and amphibians in the bestiaries and emblem books. By the time of Belon and Gesner, these animals were more than symbols; they were objects of curiosity and study for the purpose of generating new knowledge about nature. Although historical knowledge was conveyed along with contemporary natural history within the text, both Belon and Gesner relied primarily on first-hand knowledge for most of their imagery (with the curious exception of the mythical creatures).

Gesner's efforts were followed by those of the naturalist and prolific collector Ulisse Aldrovandi (Italian, 1522-1605). Aldrovandi published on many subjects, and like Gesner he produced encyclopedias of known and mythical animals.[21] Aldrovandi did not draw but employed a variety of artists. Some images were made from animals in his own collection, but many were copied from Gesner, Belon, and others. For example, in his entry on chameleons he depicts four different types. One chameleon resembles that of Belon and Gesner, but is posed differently and is not on a branch. A more interesting addition is a drawing of a chameleon, based on an animal that was sent to him, in which he shows the lizard catching a fly with its tongue.[22] This may be the earliest published image of a chameleon feeding, and it may have been meant to refute the

---

[21] Ulisse Ambrosini Bartolommeo Aldrovandi, *Vlyssis Aldrovandi Patricii Bononiensis De Quadrupedibus Digitatis Viviparis Libri Tres, Et De Quadrupedibus Digitatis Oviparis Libri Duo.* Bologna: Nicolaum Tebaldinum, 1637) and Ulisse Aldrovandi and Ambrosini Bartolommeo, *Vlyssis Aldrovandi Patricii Bononiensis Serpentum, Et Draconum Historiae Libri Duo* (Bologna: Clementem Ferronium, 1640). Both of these books were published decades after his death, and as in his other natural history books these added tremendously to the published textual information on animals (largely history, symbolism and so on). Much ink has been expended comparing Aldrovandi to Gesner and other encyclopedists, but in fact, Aldrovandi added surprisingly little in the way of images of reptiles and amphibians considering the greatly increased number of specimens to which he had access. The opposite is true for his three volumes on birds in which he added images of tropical and new world species to what had been presented by Gesner.

[22] Ibid., 670-671.

longstanding myth that these animals exist on air.[23] Aldrovandi's separate volume on serpents and dragons is notable mostly for the fact that it contains images of mythical animals such as the hydra, basilisk and various dragons in numbers roughly equal to those of actual snakes. Many of the mythical animals and some of the snakes are copied from Gesner's woodcuts, but some specific snake images have been added. These are not very well drawn, and the accompanying illustrations of internal anatomy are even poorer; for instance, the snake "skeleton" looks like a serpentine piece of spiky ribbon and bears no resemblance to a vertebral column of any kind.[24] Another woodcut shows a viper accompanied by a lateral detail of its head with the mouth open to show its teeth; however, the teeth as depicted are not those of a viper or of any other snake.[25] Considering the care taken with the images in the books published before Aldrovandi's death (i.e., the beautifully illustrated volume on birds), I must conclude that the images in these later volumes would not have met his exacting standards.[26] Gesner and Aldrovandi were followed by other encyclopedists including Jan Jonston (Polish, 1603-1675) and Edward Topsell (English, 1572-1625),[27] both of whom copied most of their images from predecessors, including

---

[23]Ashworth Jr., "Marcus Gheeraerts and the Aesopic Connection in Seventeenth-Century Scientific Illustration," 134-37. Ashworth traces the image of Belon's chameleon through the encyclopedias, emblem books and even Aesopic tales of the 16th and 17th centuries.

[24] Aldrovandi and Bartolommeo, *Vlyssis Aldrovandi Patricii Bononiensis Serpentum, Et Draconum Historiae Libri Duo*, 273.

[25] Ibid., 116

[26] Aldrovandi bequeathed his valuable natural history collection to the city of Bologna on the condition that his work was published after his death, which is the case for eight of the twelve natural history volumes published in his name. As keeper of the collection for twenty-five years beginning in 1632, Bartolomeo Ambrosini was charged with the task of seeing the books on reptiles into publication. Paula Findlen, *Possessing Nature: Museums, Collecting, and Scientific Culture in Early Modern Italy* (University of California Press, 1994), 25.

[27] Joannes Jonstonus, *Historiae Naturalis*, 6 pts. in 2 vols. (Amstelodami: Apud Ioannem Iacobi fil. Schipper, 1657) and Edward Topsell, *The History of Four-Footed Beasts and Serpents: Describing at Large Their True and Lively Figure, Their Several Names, Conditions, Kinds, Virtues (Both Natural and Medicinal), Countries of Their Breed, Their Love and Hatred to*

the mythical reptiles. Topsell's startling frontspiece to the volume on serpents (volume two) is his most original image, but the engraving of the giant arrow-tailed "boas" in the process of consuming an infant did little to advance the knowledge of reptiles.

By the end of the sixteenth century the tremendous expansion in trade and exploration provided a rich bounty of animals to be drawn "from life." Both exotic specimens and images of these strange organisms were avidly collected by a variety of people such as naturalists, apothecaries, and merchants. Many had their collections or "curiosity cabinets" documented in drawings, paintings and even published books. One such was the Italian apothecary, Ferrante Imperato (1550-1631), who maintained his collection in Naples not only as a place to conserve the raw ingredients for medicines,[28] but also as a place of study that was visited by many scholars of the day.[29] Imperato published a book describing his collection, and this volume was considered important enough to have been reprinted for over seventy years.[30] An often reproduced image from a fold-out page in the front of this volume shows visitors admiring Imperato's "cabinet" that was filled with specimens displayed in beautiful wall cabinetry and arrayed across the arched ceiling. The fact that a large crocodilian took pride of place in the center of the ceiling and several reptiles and amphibians were exhibited nearby

---

Mankind, and the Wonderful Work of God in Their Creation, Preservation, and Destruction (London: Printed by E. Cotes, for G. Sawbridge ..., T. Williams ..., and T. Johnson ... 1658).

[28] Guiseppe Olmi. "From the marvelous to the commonplace: notes on natural history museums," in Renato G. Mazzolini, ed. *Non-Verbal Communication in Science Prior to 1900* (Firenze: L.S. Olschki, 1993), 248.

[29] Frequent visitors included members of the Accademia dei Lincei of which Imperato was a member. The academy of the "Lynx-eyed" was determined to supplant classical knowledge with that gained by direct observation. Their influence on later natural history scholars was blunted by the dissolution of the group after the death of the founder, Frederico Cesi. Findlen, *Possessing Nature: Museums, Collecting, and Scientific Culture in Early Modern Italy*, 31-33.

[30] Ferrante Imperato, *Dell'historia Naturale* (Napoli: C. Vitale, 1599) was the first edition of this book.

argues for the interest these animals held for scholars of the day.[31] Images also played a key role in collections like those of Imperato, clarifying the appearance of the organism in life or at least when specimens were fresh.[32] Specimens became degraded with time and handling, and an accompanying "paper museum" helped both to defray this loss and to facilitate information exchange with other naturalists through the duplication of images. Imperato's book provides a good example of such documentation of a collection; most of the engravings appear to have been made with reference to living or fresh specimens and are not copies from other books. Three different animals are shown with their young: the "fire" salamander (*Salamandra salamandra*); a live-bearing skink, and a viper (Figure 4).[33] These early images of reptiles and amphibians with their offspring show an interest in the life cycle of the animals, and the viper illustration may have been meant to put to rest the myth that young vipers kill their mother at birth by eating their way out of the womb. The image depicts one young viper and its mother at the moment of birth, but it also shows the earlier step-by-step process that occurs as young vipers emerge from their individual amniotic birth sacs. This progression of events could have been seen only upon dissecting a pregnant viper (as with mammals the sac is broken open before the young enter the birth canal). It was essential that the original drawing of these animals be made from live or very fresh specimens, because the sacs would have become shriveled and clouded in preservative and information lost. The enlightening image of the vipers in

---

[31] Ibid. At least a dozen reptiles and amphibians in addition to the crocodile can be seen, including a frog or toad and a salamander, along with several snakes and lizards. The crocodile must have been a particular favorite (or perhaps an early form of advertising logo) because it also adorns the title page of the first edition.

[32] Guiseppe Olmi. "From the marvelous to the commonplace: notes on natural history museums." In Mazzolini, ed., *Non-Verbal Communication in Science Prior to 1900*, 240-41.

[33] Imperato, *Dell'historia Naturale*, 785-90. The image of the salamander and young omits the egg and larval stages. The skinks and snakes are very well-drawn, but without any indication of size, color or geographic origin, it is not possible to identify them further. The same is true of the viper.

Figure 4 emphasizes both the importance of direct observation and the necessity of recording such observations from live animals.

Conversely, the engraving of native Floridians hunting and killing an alligator (Figure 5)[34] appears to be more drama than fact – at least regarding the representation of the animals - and points to the shortcomings of delayed or second-hand information turned into image. The original source for this volume from de Bry's series of "Great Voyages" was Jacques le Moyne de Morgue (French, 1533-1588), one of the few survivors of an ill-fated attempt to establish a Huguenot outpost in Florida.[35] Le Moyne was a fine artist whose surviving images of plants and animals are detailed and correct, but in escaping the settlement with his life, he may have left behind his notes and drawings, because all but one of the originals has been lost. It is impossible to know how much the distortion of the alligators' appearance is the result of Le Moyne being forced to produce these drawings from memory and how much is the result of liberties taken by de Bry's engravers.[36] Other images in the same volume show animals more correctly, so it is difficult to understand why the alligators have a head shape like those shown in Figure 5 (i.e., the bumpy snout, mammalian cranium and external ears). The alarming size of the foreground animal may be exaggerated, but alligators may have been larger and humans smaller 450 years

---

[34] Theodor de Bry, *Brevis Narratio Eorvm Qvæ in Florida Americæ Provicia Gallis Acciderunt, Secunda in Illam Nauigatione, Duce Renato De Laudoniere ... Anno Mdlxiiii. Qvae Est Secvnda Pars Americae... Auctore Iacobo Le Moyne, Cui Cognomen De Morgues* (Frankfurt: I.Wecheli, 1591), plate 26. Le Moyne refers to the animals as crocodiles, but their size and setting indicate that they probably were what are now classified as alligators.

[35] Le Moyne accompanied the 1562 expedition as the recording artist. The Florida venture failed within two years due to internal strife among the Frenchmen. Paul Hulton, *The Work of Jacques Le Moyne De Morgues: A Huguenot Artist in France, Florida, and England*, 2 vols. (London: British Museum Publications, 1977), 4-10.

[36] Other images in publications by de Bry have similar exaggerations and errors in depictions of new world animals. For a discussion of this and examples of such images see Miguel Asuá and Roger French, *A New World of Animals: Early Modern Europeans on the Creatures of Iberian America* (Aldershot, England: Ashgate, 2005), 125-129.

ago.[37] Whatever the reason for the manner of depiction, the image of the alligator hunt and others within de Bry's "Voyages" served to introduce readers to an exotic and dangerous new world. Depictions of the fauna of the Americas and West Indies were critical to the process of documenting new organisms that had no history in Europe, and hence, no mythology or readily available images to be copied. One of the earliest volumes of the new world to catalog and portray a wide variety of organisms was the natural history of Brazil.[38] The simple but accurate woodcuts are presented with text describing the organism, and included are a tupinambis (lizard), a boa, and a toad, all of which are identifiable to family level from their images. Considering the abundant array of herpetofauna in Brazil (over 1300 species of reptiles and amphibians), the small selection included in *Historia Naturalis Brasiliae* is more notable for what omitted. However, these images were clearly the result of direct observation of the animals; even more of these visual observations were recorded as unpublished watercolors and sketches by artists posted to the Dutch settlement in Brazil.[39]

With the exceptions of Gesner and Belon, most of the images of amphibians and reptiles discussed to this point were created by artists rather than naturalists. When the person making observations about an organism is not the same person representing it visually, critical information may be lost. On the

---

[37] Reptiles demonstrate indeterminate growth, and like trees, continue to increase in size as they age. The contemporary record for alligator length is nineteen feet, which occurred in a period with much more hunting pressure and habitat restriction than in the sixteenth century. The alligator in Figure 5 is four to five times as long as the height of a hunter; if the man was five foot five inches tall, the alligator would be twenty-two to twenty-seven feet long, which is not unimaginable for that time. Also in Le Moyne's defense, I can state from field experience that alligators do seem exaggerated in size when encountered up close.

[38] Willem Piso and Georg Marggraf, *Historia Naturalis Brasiliae* (Amsterdam: Franciscum Hackium, 1648). This book has 429 woodcuts and contains Piso's contribution on Brazilian medicine and Markgraf's natural history. Asuá and French, *A New World of Animals*, 116-17.

[39] In the second quarter of the seventeenth century hundreds of images of plants and animals were produced by the six or more artists working for the governor-general, Johan Maurits. Many of these were sold by Maurits when he returned to Holland and thus the images disseminated throughout Europe. Ibid., 117.

other hand, a naturalist with poor artistic skills may produce an even less successful representation. The images in Oliger Jacobaeus's (Danish, 1650-1701) work on frogs and lizards exemplify this situation.[40] His illustrations of the internal anatomy of a frog and that of a lizard are fairly accurate if crude, but his simplified drawings of adult frogs in the plate on metamorphosis are strangely like the frogs of Medieval bestiaries, exhibiting disproportionately small heads, a definite "neck" and an incorrect number of toes. The tadpoles would be unrecognizable as frog larvae if seen out of context. However, Jacobaeus is of interest because he was one of the first to illustrate the stages of amphibian metamorphosis from egg to adult.

The quintessential artist-naturalist working at the end of the seventeenth century was Maria Sibylla Merian (German 1647-1717). The daughter and step-daughter of prominent engravers, Merian was trained as an artist from an early age. Her lifelong pursuits in natural history also began in childhood, and she wrote of raising silk worms to metamorphosis when she was thirteen.[41] Merian established her reputation by publishing beautifully illustrated books both on flowers and on her work with the life cycles of European butterflies and moths. Her *magnum opus* resulted from two years of work done in Dutch Surinam, and when *Metamorphosis* was published in 1705 it set a new standard for natural history books.[42] Aside from the magnificent quality of the artwork in this folio

---

[40] Oliger Jacobaeus, *De Ranis Et Lacertis Observationes* (Hafniae: Impensis Johannis M. Lieben, 1686). This volume was first published in 1676, but I had access to a later edition. Plate I shows thirteen stages of development of a frog, depicting three stages as an egg, six as a legless tadpoles, and four more stages as the legs appear and the tail is lost. These are little more than outlines with some cross-hatching, and four stages of the tadpoles are indistinguishable from each other except for increasing size.

[41] Maria Sibylla Merian, *Schmetterlinge, Käfer Und Andere Insekten: Leningrader Studienbuch*, Wolf Beer, ed. (Luzern: Reich, 1976), 51. This facsimile publication of Merian's journal or "studienbuch" contains photos of many of her original studies of insects and other animals as well as the text of her accompanying notes.

[42] Maria Sibylla Merian, *Metamorphosis Insectorum Surinamensium* (Amsterdam: Published by the author, 1705). The plates exhibit the fine detail made possible by copper-plate engraving, and in the first edition the information on the organisms was significantly enhanced by careful hand-coloring.

volume, it was the first book published by a European artist-naturalist who voyaged to the New World to study a specific aspect of nature. All or most of the organisms portrayed within were painted "from life," and many were kept alive for several weeks by Merian as she studied them.[43] As she wrote to James Petiver, she was not interested in merely cataloguing exotic specimens, but in the "formation, propagation, and metamorphosis of creatures, how one emerges from the other, the nature of their diet."[44] The emphasis on the life cycles and interactions of the species depicted makes *Metamorphosis* one of the first ecological studies (long before the word was coined). The focus of this volume and most subsequent research on it has been the detailed portrayal of the plants and insects of Surinam, while the fine images of reptiles and amphibians contained within have been largely overlooked. The first edition of *Metamorphosis* included two types each of lizards, snakes, and frogs, all painted precisely enough to be identifiable to species.[45] What is remarkable about Merian's depictions is that the "loathsome beasts" are portrayed in a way that sparks kinship and curiosity. In the plates containing the frogs she shows the adult with its young; Figure 6 depicts an adult tree frog (*Phyrnohyas venulosa*) and the stages of metamorphosis, showing the early eggs, an egg about to hatch, four stages of tadpole, and small froglet with a tail.[46] An older froglet (larger and

---

[43] Ibid., *passim.*

[44] Maria Sibylla Merian, William T. Stearn, Vitor Osmar Becker, and Elisabeth Rücker, *Metamorphosis Insectorum Surinamensium*, 2 vols. (London: Pion, 1980), 73. This facsimile volume contains commentary and a collection of Merian's letters. Merian wrote to Petiver on 27 April 1705 to tell him about the completion of her book and to ask him not to send her any more "small creatures," as she was not interested in collecting animals about which she had no background.

[45] Although Merian's purpose was not to catalog or classify, her images were so accurately rendered that they were used by others to do just this. Linnaeus named several species based upon Merian's images. William T. Stearn, "Maria Sibylla Merian (1647-1717) as a Botanical Artist," *Taxon* 31:3 (1982), 532.

[46] Images in Figures 6, 7, and 8 are photographs of original watercolors painted by Merian that may have served as models for the engravers working on *Metamorphosis*; these are

with a shorter tail) has met its fate as a meal for a predatory water bug.[47] The tableau is striking in its contrast between the beauty of the water plant and the birth-to-death struggle of the animals swimming around its base. As with Merian's other plates in *Metamorphosis* this microcosm of life shows a much more complete picture of "nature" than any illustrations of her predecessors. Merian also was the first to record the image of the remarkable *Pipa pipa* (the Surinam "toad"), an aquatic frog whose fertilized eggs become embedded in the dorsal integument where they develop until they hatch out as little froglets. She depicted this unusual phenomenon more clearly than can be achieved with a photograph, showing all of the stages in one image.[48] The Surinam "toad" was the first known example of an amphibian without a free-swimming larval stage. The biology of reproduction in amphibians restricts them to reproducing in water no matter how terrestrial the adult stage, and Merian's depictions of metamorphosis contributes to the understanding of this process.

The small lizard posed on a banana plant with its eggs and one of its hatchlings (Figure 7) likewise is typical of Merian's attention to aspects of animal reproduction,[49] as is the striking image of the caiman and the snake that appeared

---

from a complete set acquired by Hans Sloane that is now in the British Museum, but another set is housed at the Windsor Royal Library and a partial set is in St. Petersburg at the Russian Academy of Science.

[47] Merian did not know that this species is arboreal, because she found them in the water while they were breeding. She thought that the toe pads (characteristic of climbing tree frogs) were an aid to moving about in the marshy habitat, and she mistook the extended vocal sacs of the males for "ears." She did however document how long each stage lasted in the metamorphosis, as she frequently did for the caterpillars of moths and butterflies. Merian, *Metamorphosis Insectorum Surinamensium* text accompanying Plate 56.

[48] Ibid., plate 59 and text. Merian's text also describes the emergence of the young from the back of the adult. She relates preserving the adult and her young in brandy, probably in order to have a specimen to observe more closely as she painted it.

[49] The lizard is identifiable as the rainbow whiptail, *Cnemidophorus lemniscatus*. This image is reproduced in the 1705 volume of *Metamorphosis* as plate 23.

in a posthumous edition of *Metamorphosis* (Figure 8).[50] Like its much larger crocodilian cousins, the caiman will defend the eggs in its nest, and such a defense is shown here. As has been pointed out by others, the size of the false coral snake (*Anilius scyta*) seems exaggerated, but it may be that Merian used a juvenile spectacle caiman (*Caiman crocodilus*) as her model. The coloration of Merian's caiman is more similar to that of a juvenile than of an adult, and a subadult could be shorter than this species of snake, which can be a meter long. Dramatic license or no, crocodilians are unusual among reptiles in showing maternal behavior, and Merian has depicted this unexpected role of a fearsome creature succinctly. Several unpublished watercolors of other reptiles and amphibians by Merian exist in various collections, and she wrote that she hoped to publish another book on "such creatures" if the public and naturalists expressed an interest;[51] unfortunately, the time in Surinam took a toll on her health and this project was never realized.

The influence of Merian may be seen in many important natural history studies that followed the publication of *Metamorphosis*, including those of Mark Catesby (English, 1683-1749) and of August Johan Roesel von Rosenhof (German, 1705-1759). Catesby was a naturalist and professional collector who traveled extensively in what was then known as "Carolina, Florida and the Bahama Islands."[52] From 1729-1747 he worked on images and text for his two-

---

[50] Maria Sibylla Merian, *Metamorphosis Insectorum Surinamensium* (Amsterdam: Joannem Oosterwyk, 1719). In this edition twelve plates were added to the original sixty (all but two of the additions are thought to be by Merian). As with the original plates, multiple watercolor copies of some of these additional images were made. For example, I have seen the same image of the caiman and snake produced in watercolor and ink on vellum at the Morgan Pierpont Library, at the Royal Library at Windsor and at the British Museum. It is still unclear to scholars why multiple "originals" exist, but I suspect that Merian, known to be canny in business, produced them because wealthy collectors were willing to pay a premium for them.

[51]Merian, *Metamorphosis Insectorum Surinamensium,* text accompanying plate 4. This plate shows a striking lizard (a young tupinambis) on a cassava plant.

[52] Catesby's travels were supported by a number of patrons for whom he collected plants and seeds. For more on Catesby and his *Natural History* see George Frick and Raymond Stearns, *Mark Catesby: The Colonial Audubon* (Urbana: University of Illinois Press, 1961).

volume *Natural History* on the flora and fauna of this region.[53] As seen in Figure 9 of the bullfrog (*Rana catesbeiana*, named for the naturalist) and the ladyslipper plant, most of the folio-sized Catesby plates featured one species of plant and one animal. Catesby's plants were not always associated with the animal depicted and may or may not have even been found in the same habitat.[54] He covered a much wider geographic area than Merian and his *Natural History* includes over two-hundred plates of birds, mammals, reptiles, amphibians and fish. The herpetofauna represented in Volume II of Catesby's *Natural History* include three sea turtles (one depicted with her eggs buried nearby), thirteen species of snakes, six species of lizards, four frogs and a young alligator. In most cases the animal is portrayed in a predominately dorsal view with a flattened appearance. Although Catesby does seem to intend a lifelike pose for most of the animals in his book, some of the snakes are arranged in what could only be described as a decorative manner, exhibiting rather unnatural configurations of bends and coils. Catesby conveys special attention to the rattlesnake by placing it on a page without an accompanying plant. The snake is coiled in a defensive posture, rattle raised. He also includes a detailed drawing of a dissected fang and two details of detached rattles.[55] It is not surprising that the details are included for this specific snake due to its reputation for danger to humans, but the illustrative nature of this image is in clear contrast to most of his plates. A few of Catesby's plates do show some

---

[53] Mark Catesby, *The Natural History of Carolina, Florida and the Bahama Islands: Containing the Figures of Birds, Beasts, Fishes, Serpents, Insects, and Plants; Particularly the Forest-Trees, Shrubs, and Other Plants, Not Hitherto Described, or Very Incorrectly Figured by Authors. Together with Their Descriptions in English and French. To Which, Are Added Observations on the Air, Soil, and Waters; with Remarks Upon Agriculture, Grain, Pulse, Roots, Etc. To the Whole, Is Prefixed a New and Correct Map of the Countries Treated of, 2 Volumes* (London: 1731-1743).

[54] It is unclear whether he was attempting to save on printing costs by combining these figures on a page. The original volume was self-published and sold by subscription; he learned engraving to produce these. David R. Brigham, "The Patronage of Natural History," in *Empire's Nature: Mark Catesby's New World Vision*, Amy Meyers and Margaret Pritchard, eds. (Chapel Hill: University of North Carolina Press, 1998), 114-115.

[55]Catesby, plate 41.

interaction among organisms. His "brown viper" is about to prey upon a hapless salamander at the water's edge,[56] and in another plate his tree frog has its mouth agape below a dangling spider.[57] Mark Catesby's contribution was significant primarily because he presented a window on the diversity of reptiles and amphibians in the "new world;" his inventory certainly is not complete (for example, freshwater and land turtles are conspicuously absent), but he showed a greater variety of species than any predecessor. Additionally, the presentation of a bullfrog, snake or any creature arrayed across a folio-sized page, isolated from text and carefully hand-colored, imbues substance and significance to the animal. Such images assert that the organisms pictured are worthy of study and consideration.

While Catesby's volumes presented an overview of the natural history of a region, von Rosenhof's richly illustrated treatise on frogs was an exercise in detail and a bold new step in the imagery of amphibians that surely elevated their status as animals worthy of study.[58] In 1728 von Rosenhof was introduced to Merian's *Metamorphosis*, and this event is credited with inspiring his book on the natural history of German insects.[59] He followed this volume with publication of the *Historia Naturalis Ranarum*. The magnificent frontispiece of this book (Figure 10) seems a tribute to Merian; the lizard twined around the rose stem regarding a butterfly, the details of the rose plant showing all stages from bud to fruit, and the evidence of insect damage on the plants' leaves all show the hallmarks of Merian's artistic style and observational skills.[60] *Historia Naturalis Ranarum* covers the small number of German anurans with great thoroughness in both text

---

[56] Ibid., plate 45.

[57] Ibid., plate 71.

[58] Rösel von Rosenhof, *Historia Naturalis Ranarum Nostratium.*

[59] August Johann Rösel von Rosenhof, *Der Monatlich-Herausgegebenen Insecten-Belustigung* (Nuremberg: Johann Joseph Fleischmann, 1746).

[60] Rösel von Rosenhof, *Historia Naturalis Ranarum Nostratium.*

and image. The book is subdivided into sections by species, and for each type of frog or toad von Rosenhof illustrated every step of their reproductive cycle from mating through metamorphosis as well as their external and internal anatomy. Aside from some small vignettes of the animals in their habitat and the full page frontispiece, the animals are simply arrayed on a white page. The forty-eight plates in the book are arranged in pairs of duplicates; the animals are first presented in color and unlabeled, and this is followed by a plate in black & white with figure labels that are referenced in the text. This arrangement was undoubtedly designed to satisfy both the artist and the naturalist in von Rosenhof. The result was a splendidly illustrated natural history study that is considered one of the finest of its type even today.

Natural history studies of the eighteenth century were capped by publications such as the 44-volume *Histoire Naturelle* by Georges-Louis LeClerc, Comte de Buffon (French, 1707-1788).[61] Reptiles and amphibians were covered in volumes written by Bernard Germain Étienne de Lacépède, and the images in these books vary tremendously in quality. Some would be quite useless for identifying an animal beyond a broad grouping, and the plates are consistently lacking in any information about the animals' habits or habitat.[62] LeClerc and Lacépède were concerned less with such details than with the ordering of nature by classification and cataloging as much of it as possible. However, by the middle of the nineteenth century exquisitely detailed and naturalistic images of every sort of plant and animal were made available by the process of lithography. Illustrated natural history books were the rage of the publishing world as an increasingly educated public became entranced by nature, and studies on reptiles

---

[61] Georges Louis Leclerc comte de Buffon, *Histoire Naturelle, Générale Et Particulière Avec La Description Du Cabinet Du Roi*, 44 vols. (Paris: De l'Imprimerie royale, 1749-1804). Many editions of this were issued in an ever-expanding number of volumes.

[62] Many of these illustrations seem almost cartoon-like when compared to those by Merian, Catesby, and Von Rosenhof. For example, "Le Scinque," depicts a stubby lizard vaguely resembling a skink, posed in a classical setting atop the ruin of a Greek statue. Ibid., volume 37, 373.

and amphibians ranged from monographs to broad surveys.[63] By the time Charles Darwin returned from his famous voyage with specimens of three Galapagos tortoises, the study of reptiles and amphibians was an integral part of the beginnings of modern biology.[64]

Conclusions

The above chronology of selected images is not meant to be comprehensive, or to imply that there was some orderly and linear evolution of increasingly representational portraits of these animals. Indeed, Gesner's salamander was much more accurately rendered in the sixteenth century than the same species pictured in the Buffon's encyclopedia over two-hundred years later. The manner in which animals were portrayed changed in response to many factors, but the purposes served by these images are the key to understanding their context. Likewise, understanding the intended function of the images gives us insight into how these animals were perceived at various points in history. That the salamander in an early bestiary resembled no living creature was not important to the symbolic function of the illustration. The earliest naturalistic portraits of nature were those of flora and fauna used for food or medicine, because reliable identification was essential.[65] Naturalistic representation of reptiles and amphibians came much later with the work of Belon and Gesner, when the purpose of the image was to show the animal as it was observed, and to

---

[63] For a pictorial overview see Sally Haines, *Slithy Toves: Illustrated Classic Herpetological Books at the University of Kansas in Pictures and Conversations* (Ithaca: Society for the Study of Amphibians and Reptiles 2000), 68-98.

[64] Charles Darwin, *Journal of Researches into the Geology and Natural History of the Various Countries Visited by H. M. S. Beagle, under the Command of Captain Fitz Roy, R. N., from 1832-1836* (London: Henry Colburn, 1839), 394. The fact that tortoise populations isolated on different islands varied enough to be distinguished from each other was one of the observations that helped Darwin to formulate his ideas on natural selection.

[65] Manuscript herbals with images of identifiable plants are known from as early as the fifth century, and by the seventh century birds were depicted realistically. Collins, *Medieval Herbals*, 37-40.

place it in the growing catalog of nature. While the text of Gesner and Aldrovandi included information on the symbolism of the animals, with the exception of the mythical creatures pictured, the images are straightforward and naturalistic. William Ashworth has argued that Aldrovandi brought "the emblematic view of nature to fruition," but he also goes on to say that within a half century of Aldrovandi's death this view had dissipated.[66] When exotic animals from the Americas and elsewhere came to be viewed in Europe, there was no historical context or established symbolism. By the middle of the seventeenth century the purpose behind animal images shifted to cataloging and describing the natural world, and reptiles and amphibians were included in this process, albeit not as comprehensively as other fauna. Images of large or venomous reptiles did play a role in showing how alien and dangerous these new lands could be, and possibly, how exciting.[67]

Maria Sibylla Merian was among the first to make images with a purpose that was clearly separate from that of cataloging and classifying. Before Merian, reptiles and amphibians were typically pictured as specimens on a page with no external context. A few images illustrated an animal feeding or some aspect of reproduction, but little more. Merian's reptiles and amphibians (and of course her insects) are shown as existing in nature with plants and with other animals. Her animals are portrayed with their natural coloration; they exhibit behaviors, reproduce, and serve as both prey and predators. This quantity of visual information demystifies the beasts, and showcasing these images on a folio-sized page validates their significance. Mark Catesby greatly expanded the number of reptiles and amphibians given this star treatment, although the purpose of his images seems to have been aligned somewhat more with those who wished to

---

[66] William B. Ashworth, "Emblematic natural history of the Renaissance" in Nicholas Jardine, James A. Secord, and E. C. Spary, *Cultures of Natural History* (Cambridge: Cambridge University Press, 1996), 30-36.

[67] This seems to be the purpose behind the image in Figure 5.

inventory nature. However, like Merian and Von Rosenhof, Catesby granted a certain aesthetic quality to the herpetofauna he drew and painted, and it is interesting to note that he usually reserved the more sinister poses (coiled body, open mouth) for venomous species of snakes such as the rattlesnake and brown viper. Non-poisonous snakes like the green tree snake often are draped around flowering plants like baroque ornaments.[68] It may be that his purpose in this presentational style was to emphasize their harmless character as well as their physical beauty.

Lastly, I would like to discuss the importance of the images of amphibian metamorphosis that began in a simple way with Gesner and reached their apogee with von Rosenhof. Depictions of this gradual process of development from egg to adult body form were important for several reasons. Perhaps most obviously, these images show the process of embryonic development, which is not so visible in most animals. Metamorphosis also emphasizes that animals are not immutable, but can change as they grow. Unique to amphibian metamorphosis is the transformation from a creature that breathes in water to one that can respire in air. Finally, images of the spawning of eggs and their development helped to dispel long-held beliefs in spontaneous generation. All of these aspects of the study of metamorphosis in amphibians made an important contribution to what is now modern biology.

In observing nature directly and recording these observations in text and images, natural philosophers took the first steps to establish scientific inquiry, pushing aside the role of myth and magic in understanding the world. Curiosity about the "loathsome beasts" and increasing attention to their natural history reflect this sometimes circuitous path very well. Even though some naturalists in the Early Modern period treated these animals with the same objective attention as the other vertebrates, others such as naturalists John Ray and Thomas Pennant

---

[68] Catesby, Plate 47.

expressed a deep aversion to them.[69] Such repugnance and fear may actually be explainable to a degree by twenty-first century science. Psychologists working with photographs of snakes found that humans and other primates learn fear of snakes more quickly than fear of almost any other stimulus, and they have postulated that there may be an evolutionary reason for this.[70] It appears that images of the loathsome creatures may teach us about ourselves as well as about the nature of the beasts.

Kay Etheridge (Gettysburg College) is Associate Professor of Biology at Gettysburg College. Her publications include work on foraging in tropical bats, food consumption by manatees, and the role of thyroid hormones in the regulation of metabolism in reptiles and amphibians. She currently is studying the evolution of images in natural history studies from the medieval through the early modern periods.

---

[69] Keith Thomas, *Man and the Natural World: Changing Attitudes in England, 1500-1800* (New York: Oxford University Press, 1996), 57. Ray, an eminent seventeenth century English naturalist, stated that he had such an abhorrence of snakes that he was "not very inquisitive after them." Pennant, an eighteenth century Welsh naturalist who published on the quadrapeds, admitted a dislike of frogs that prohibited his examination of them.

[70] Arne Öhman and Susan Mineka, "The Malicious Serpent: Snakes as a Prototypical Stimulus for an Evolved Module of Fear," *Current Directions in Psychological Science* 12:1 (2003), 1-5. It can be argued than an innate fear of a potential danger is a protective adaptation that would be selected for in offspring.

Figure 1. Salamander. From the Aberdeen Bestiary, English 1200. Aberdeen University Library MS 24, folio 70r. Used with permission of Aberdeen University. Copyright University of Aberdeen.

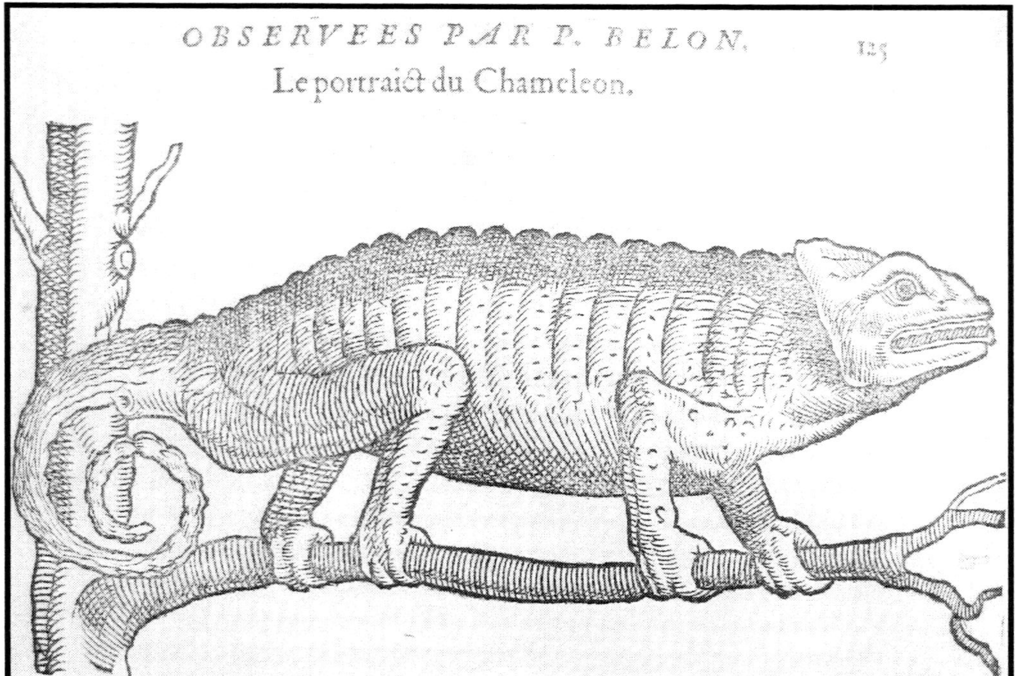

Figure 2. Chameleon. From Belon, Pierre. *Les Observations De Plusieurs Singularités Et Choses Mémorables, Trouvées En Grèce, Turquie, Judée, Egypte, Arabie & Autres Pays Étranges.* Paris: Chez Guillaume Cauellat, 1554. Image courtesy of the Department of Library Services, American Museum of Natural History.

Figure 3. Mythical and real salamanders. From Gesner, Konrad. *Historia Animalium Liber II: De Quadrupedibus Ouiparis.* Frankfurt: Roberti Cambieri, 1586. Used with permission of the Smithsonian Institution Libraries, Washington, DC.

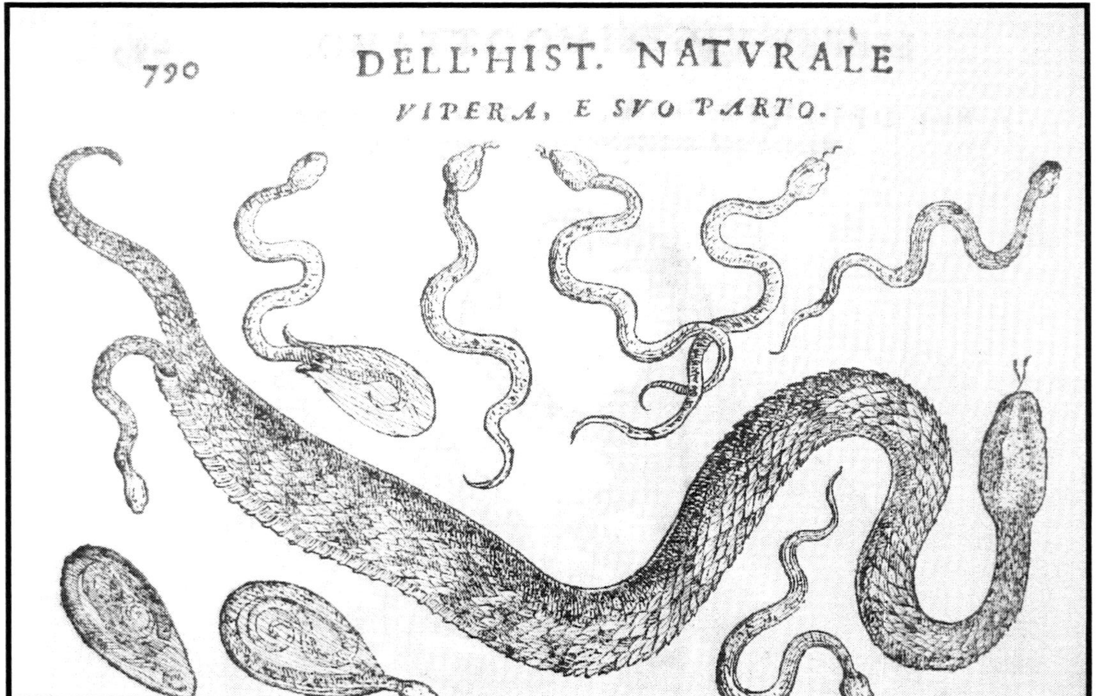

Figure 4. Viper giving birth. From Imperato, Ferrante. *Dell'historia Naturale*. Napoli: C. Vitale, 1599. Image courtesy of the Department of Library Services, American Museum of Natural History.

Figure 5. "Their way of killing crocodiles" by Jacques Le Moyne de Morgue. From de Bry, Theodor. *Brevis Narratio Eorvm Qvæ in Florida Americæ Provicia Gallis Acciderunt, Secunda in Illam Nauigatione, Duce Renato De Laudoniere ... Anno Mdlxiiii. Qvae Est Secvnda Pars Americae... Auctore Iacobo Le Moyne, Cui Cognomen De Morgues.* Franfurt: I.Wecheli, 1591. Image courtesy of the Department of Library Services, American Museum of Natural History.

Figure 6. Frog with tadpoles and eggs, hyacinth and insects. Maria Sibylla Merian. Pen and ink with watercolor and body-color on vellum.

Figure 7. Lizard with eggs and hatchling, butterflies and banana plant. Maria Sibylla Merian. Pen and ink with watercolor and bodycolor on vellum.

Figure 8. Caiman defending young and egg against a false coral snake. Maria Sibylla Merian. Pen and ink with watercolor and bodycolor on vellum.

Figure 9. Bullfrog and ladyslipper. Mark Catesby. *The Natural History of Carolina, Florida and the Bahama Islands*: London, Printed at the expense of the author, 1731-1743.

Image courtesy of the Department of Library Services, American Museum of Natural History

Figure 10. Frontispiece from Rösel von Rosenhof, August Johann. *Historia Naturalis Ranarum Nostratium*. Nuremberg: Johann Joseph Fleischmann, 1758.

Photo by David R. Patterson, used with permission.

Figure 11. Hardwick Hall, Derbyshire, 1590-1597.

Photo by the author.

Figure 12. Wollaton Hall, Nottingham, 1580-1588.

Technocrat of Gunpowder and Cannon: Satan's Mechanical Contrivances in *Paradise Lost's* War in Heaven

Elizabeth Oldman

My study derives from revolutionary advancements in military technology which molded the modern nation-state throughout Renaissance Europe. Gunpowder fabrication and correlated weaponry were essential aspects of the radical innovations which took place at this time. As the first invented chemical explosive, gunpowder was an offshoot of Chinese alchemy and its engagement with nitrate compounds.[1] The route via which the West gained knowledge of the compound remains uncertain, but it is believed that the transmission must have occurred during the thirteenth century, when Christendom, Islam, and the Orient widely exchanged ideas and goods. Interestingly, Europeans immediately used gunpowder in the "most bellicose of its applications,"[2] in order to improve upon weapons for battle. As Hall confirms, "1300--1600 . . . is not only the age often referred to as the Renaissance but also the period in which Western Europeans became accustomed to the firearm as a conventional feature of warfare."[3] Following the implementation of this new technology, efforts to enhance firearms and to defend against assault by these weapons became widespread. Gunpowder weaponry has a momentous influence upon the costs of armed strife as tactics, equipment, and modes of fortification needed to be updated.

---

[1] See Bert S. Hall, *Weapons and Warfare in Renaissance Europe: Gunpowder, Technology, and Tactics* (Baltimore & London: The Johns Hopkins University Press, 1997), 41. Hall argues that the earliest-known recipe for gunpowder may be traced to A.D. 1044, during the Sung Dynasty in China, but the basic ingredients were by that time at least one hundred years old. The majority of scholars believe that gunpowder dates back to the ninth century in East Asia, where it was used for firework manufacture long before its potential in martial warfare was realized.

[2] Ibid.

[3] Ibid., 6.

Although associated with a previously-unforeseen destructive potential, and commonly believed to have been contrived by the devil himself,[4] firearms were consistent with the positive attitudes toward innovation which characterized the Renaissance. "The growing scientific interests of the moderns welcomed the gun for what it revealed about ballistics and the laws of motion, and what its technology taught about metallurgy and the chemistry of explosives"[5] Pervasive interest in martial discord as a subject of study began in this period. J.R. Hale further confirms that during the fifteenth and especially sixteenth centuries:

> Wars became more expensive, touched more pockets, inducted more radical changes in administrative institutions and in constitutional practice . . . [Battle], together with the individual's relationship to it, was brought into . . . clearer focus by more demanding notions of citizenship and nationhood.[6]

Armed strife gained prominence in political treatises and impacted upon international law to such a degree that it developed into a new discipline. Issues of rightful versus criminal battle, ethical leadership, and call for *temperamenta*, or moderation, in war were considered under the rubric of legal investigation.

In the analysis of Milton's representation of war which follows--and unlike the majority of critics who champion the portrayal of a pacifistic Milton--I seek to demonstrate that the poet's belief in the possibilities of limited combat were supported by an extensive body of literature and philosophy which shaped his society's views about this practice. Most specifically, I wish to argue, the individual known as the father and founder of international law, the Dutch scholar Hugo Grotius, is key to understanding Milton's view of armed strife, and is

---

[4] Jack Kelly elaborates upon this association in his prologue to *Gunpowder--Alchemy, Bombards, and Pyrotechnics* (New York: Basic Books, 2004): "Early in its history, gunpowder was labeled the "devil's distillate." Onlookers were terrified by its flash and boom. Its fabricators were secretive, blackened men, daredevils whose arcane work was subject to disastrous accidents. One of gunpowder's ingredients, brimstone, was the burning stone always associated with Satan. Gunpowder's action was a diabolical mystery--once ignited, it blazed wildly, infernally, leaving behind the sharp tang of sulfur and a haze of smoke" (ix).

[5] J.R. Hale, *Renaissance War Studies* (London: The Hambledon Press, 1983), 405.

[6] Ibid., 335.

essential to understanding why the poet would choose to portray the War in Heaven in complexly-wrought realistic detail.

Milton's concept of battle, as it was portrayed his prose and poetry, was strongly informed by natural law as it was elucidated in the Dutch scholar's works. As a foremost proponent of just / criminal war theory during the sixteenth and seventeenth centuries, Grotius envisioned the legitimate use of martial means as an instrument of rational, civilized men, employed to protect human society. Deriving his notion of the state of nature from Aristotelian and Stoic tradition, he believed that individuals are impelled by the understanding that society functions in harmony with mortal nature. As we will look at more closely, the Dutch scholar argued that this position of sociability serves as the basis of all law--natural law, originating from God, and civil law, arising from man-made legislation--which derives from society. Maintaining that the separation between the law of nature and its materialization into positive law could remain extremely narrow, Grotius contended that man-made law could elucidate the laws of nature--that legal positivism could inspire an even truer or more "natural" natural law--more naturally progressing in accordance with human experience. In this way, he surmised that exemplary standards of justice could be integrated into the humanly-fabricated rules by which individuals lived their lives.

Milton's famous meeting with Grotius took place during the poet's visit to Paris during 1638-9. He "ardently desired" (*Second Defense*, 4: 615)[7] this encounter, as interest in the Dutch scholar's sociability theories was rapidly expanding throughout England at that time.[8] In his *Second Defense*, the poet recalls:

---

[7] The prose of John Milton is quoted from *Complete Prose Works* of John Milton. 8 vols. Ed. Don M. Wolfe et al. (New Haven: Yale University Press, 1953-83), and is cited within the text.

[8] More than any other single scholar, Hugh Trevor-Roper has researched England's key role in Grotius's biography. See in particular his *Catholics, Anglicans, and Puritans, Seventeenth-Century Essays* (Chicago: University of Chicago Press, 1988).

> From the recommendation of others, I was received at Paris with The utmost courtesy, by the noble Thomas Scudamore, Viscount of Sligo, who of his own accord introduced me, accompanied by several of his suite, to the learned Hugo Grotius, at that time ambassador from the queen of Sweden to the king of France, and whom I was very desirous of seeing. (8: 122-23)

J.M. Evans confirms that "In May or June of 1638 Milton visited the Great Dutchman in Paris on his way to Italy. Soon after his return to England he began work on the dramatic drafts of *Paradise Lost,* of which the four versions preserved in the Trinity MS. date roughly from 1639 to 1642."[9] The poet refers to Grotius in passages of his works which promote natural-law derived theories regarding individual integrity and the supremacy of human reason such as *The Doctrine and Discipline of Divorce* (1643), *The Judgement of Martin Bucer* (1644), and the conclusion of *Tetrachordon* (1645).[10] Affirming his desire for easy access to Grotian philosophy and poetry, Milton's personal library contained an extensive assortment of the Dutch scholar's works, including *Adamus Exul,*[11] *Ad Genesin, Annotationes Ad Veterum Testamentum, Annotationes in Libros Evangeliorum, Christus Patiens, De Veritate Religionis Christianae*, and *Poemata Collecta,* in addition to his foremost *De Jure Belli ac Paci.*[12]

---

[9] J. Martin Evans, *Paradise Lost and the Genesis Tradition* (Oxford: Clarendon Press, 1968), 212.

[10] In regard to Grotius's notes on the law which addresses marital termination in *Annotationes in Libros Evangeliorum* (1641) on Matthew 5, Milton references the Dutch scholar in three works: in *The Doctrine and Discipline of Divorce* (1643), where he speaks of him as "*Hugo Grotius*, a man of these times, one of the best learned" (2: 238); in *The Judgement of Martin Bucer* (1644) (2: 433); and in *Tetrachordon* (1645), where the poet prefaces his quote from Grotius with "*Grotius* yet living, one of prime note among learned men" (2: 715).

[11] Grotius composed this five-act classical tragedy on the fall of man when he was only eighteen years old. The extent to which this drama serves as an essential structural model for Milton's epic has been debated.

[12] See Jackson Campbell Boswell, *Milton's Library: A Catalogue of The Remains of John Milton's Library and An Annotated Reconstruction of Milton's Library and Ancillary Readings* (New York and London: Garland Publishing, 1975), viii.

A consideration of prior history is critical to understanding the incentive behind and opportunity for Grotius's task in *De Jure Belli ac Paci.* A state of lawlessness in regards to martial strife prevailed during the Dutch scholar's time, and shaped his concept of the restraints that natural law placed upon it. Though battle was perceived as an intrinsic aspect of the international community beginning in the Renaissance, it achieved this status only in an inchoate and unexplicated manner. The rudimentary nature of martial theory, and its lethal consequences, served as the driving force behind Grotius's reason to write *De Jure Belli ac Paci* as he perceived it. He sought to address the problem of the incompatibility between developing theory and actual practice in relation to armed strife. In *The Causes of War*, Michael Howard confirms that "the problem of the control and legitimization of military power has been a central concern of writers on international relations since Grotius wrote his *De Jure* at the beginning of the seventeenth century."[13] In particular, this dilemma ignored the legitimate parameters of war. Wrongful handling of prisoners, random killing of women, children, plowmen, and peasants, were frequent battle-related occurrences during this period. Hedley Bull remarks that:

> the canonists and the schoolmen said little about the humane treatment of enemies in warfare . . . the paramount issue was the right to wage war . . . In the early Middle Ages . . . these ideas did little to abate the fury and brutality of medieval warfare, and may indeed have aggravated it. In general the doctrinal aspects of the 'just war' theory had little or no effect on the conduct of war.[14]

The theoretical tenets of evolving "just war" theory had little reflection in the barbaric and unbridled aspect of military practice.

On a more practical level, unrestricted warfare prevailed for Grotius to witness during his own era in the interminable, all-destructive fury[15] of the Thirty

---

[13] Michael Howard, *The Causes of War* (Cambridge: Harvard University Press), 39.

[14] See Hedley Bull, Benedict Kingsbury, Adam Roberts, eds., *Hugo Grotius and International Relations* (Oxford: Clarendon Press: 1990), 184.

[15] See Geoffrey Parker, *The Thirty Years War* (London and Boston: Routledge and Kegan Paul, 1984), 210.

Years War, a series of battles fought between 1618 and 1648 by various nations for myriad reasons, including religious, dynastic, territorial, and commercial values. His political works were overwhelmingly inspired by these tempestuous times. He sought to denounce the brutalities of this conflict in hope of remedying harsh military conduct. From even the incipient stages of this battle, "[c]ommanders were unable to impose military discipline of the type which would be taken for granted in the wars of the eighteenth century."[16] The Dutch scholar's *De Jure Belli ac Paci*, written at the beginning of the apex of the Thirty Years War in 1625, sought to demonstrate that the unruly behavior of soldiers contradicted the laws of nature. Into the 1630s and beyond, Grotius foresaw, increasing shortages of funding and provisions would incite military misbehavior on an even more massive scale.

Most historians deny that there was any single purpose behind or tangible result to these series of battles. At the conclusion of her classic study *The Thirty Years' War* (1938), British historian C.V. Wedgwood states that "The war solved no problem. Its effects, both immediate and indirect, were either negative or disastrous. Morally subversive, economically destructive, socially degrading, confused in its causes, devious in its course, futile in its results, it is the outstanding example in European history of meaningless conflict."[17] The final ten years of this protracted discord, in particular, saw rival armies tramping back and forth over Germany and splintering off mercenary bands as they traveled. The line between soldier and brigand became exceptionally thin by the 1640s, as control of armies became more difficult to achieve. Unable to collect their pay, mercenaries were forced out upon the countryside in order to gain supplies, thus instigating the "wolf-strategy" that characterized this conflict.

---

[16] Stephen J. Lee, *The Thirty Years War* (London: Routledge Ltd, 1991), 53.

[17] See C.V. Wedgwood, *The Thirty Years' War* (London and New York: Methuen, 1981), 526.

The armies of both sides plundered as they proceeded onwards, leaving cities, towns, villages, and farms ruined. In *The Thirty Years War*, Geoffrey Parker explains that "everywhere the records describe frequent brutality by soldiers, sporadic resistance by peasants, cautious compliance by townspeople, and desperate efforts by civilian and military officials to maintain minimum standards of justice and prevent a total collapse of law and discipline."[18] Ronald G. Asch further observes that:

> Lack of supplies and of cash was . . . a problem for all armies, and in fact, it was this problem which was to a large extent responsible for the undisciplined behavior of soldiers. At least from the later 1630s onwards, starved or half-starved soldiers were quite a common sight, and undoubtedly the desperate state of this *soldatesca* was largely responsible for the atrocities which they committed.[19]

The devastating campaigns of the Thirty Years War occurred over most of Europe, and when they ended with the Treaty of Westphalia in 1648, population displacement as well as material and economic decline consumed Germany--in particular--for decades to come. Early estimates, confirm Parker, suggested that this large-scale battle had managed to destroy half or even two-thirds of the German population.[20]

The transition from medieval to modern warfare, influenced by the rise of capitalism and the emergence of the nation-state, instigated organization of a standing professional army and dramatic technological advancements. Nevertheless, even as the shape of warfare drastically changed from the Middle Ages to the Renaissance, there was insufficient codification and regulation in response to expansion, as analyses of the Thirty Years' War and other martial scenarios prove. Although armed strife became more public as well as amplified

---

[18] Parker, 210.

[19] Ronald G. Asch, *The Thirty Year's War: The Holy Roman Empire and Europe, 1618-48* (New York: St. Martin's Press, 1997), 177.

[20] See Parker, 210.

in magnitude into the sixteenth century, cruel and gratuitously destructive practice persisted. Edwards explains that:

> Wars, from the sixteenth century onward, became international in the sense that they were fought under the banners of the new monarchs. These conflicts were broad in scope and replaced the 'private' wars of the feudalistic era . . .Unfortunately, whatever practices or customs that had found acceptance over time in regard to intersocietal relationships came to be minimized or even ignored entirely. This growing disregard for any restraint on state behavior led monarchs to adopt Machiavellianism either by design or by default.[21]

Abhorring the Machiavellian predilection for deviousness and politics divorced from ethics, the Dutch scholar railed against the discord which prevailed throughout Europe and the mounting disregard by sovereigns of policies prioritizing principles of justice and morality.

Grotius's criticism of inhumanity and lack of restraint in war-making has been widely regarded. In *Prolegomena* 28, he expressed his opposition to the anarchy which characterized both the martial conflicts of history and his own time:

> Throughout the Christian world I observed a lack of restraint in relation to war, such as even barbarous races should be ashamed of; I observed that men rush to arms for slight causes, or no cause at all, and that when arms have once been taken up there is no longer any respect for law, divine or human; it is as if, in accordance with a general decree, frenzy had openly been let loose for the committing of all crimes. (21)[22]

The Dutch scholar's ethical and legal standards were unremittingly challenged by the persistence of what he characterized as criminal wars. He was distressed by the frequent propensity for those in authority to sanction battle practices which overtly contradicted the basic tenets of natural law.

---

[21] Charles S. Edwards, *Hugo Grotius, The Miracle of Holland: A Study in Political and Legal Thought* (Chicago: Nelson-Hall, 1981), 84.

[22] All quotations from Hugo Grotius are taken from *The Law of War and Peace (De Jure Belli ac Paci)*, trans. Francis W. Kelsey (Indianapolis: Bobbs-Merrill Co., Inc., 1925), and are cited within the text.

Interestingly, Grotius's efforts to define and prohibit unlawful behavior in war and to designate an international society founded on the tenets of natural law *De Jure Belli ac Paci* coincided with a lapse in technological progress, including the fabrication of gunpowder-related weaponry. The unforeseen scale of atrocities of this protracted battle, it seems, served as a grim reality indicator. In *Gunpowder: Alchemy, Bombards, and Pyrotechnics*, Jack Kelly reveals that:

> a curious stasis . . . gripped military technology from the end of the Thirty Years War in 1648 until well into the 1800s. In spite of the fact that men of science were gradually gaining an understanding of the rules that governed gunpowder's actions, military leaders did not seek out ways to increase its effectiveness. The development of gunpowder weaponry came to a virtual standstill.[23]

This resistance by governments and military leaders to further modernize martial weaponry represented an acknowledgement that military strife had become entirely too barbarous.

Serving as a formal response to these unbridled martial practices, *De Jure Belli ac Paci* (1625) was one of the first modern attempts to impose codes on the conduct of war and use of technology in battle. In this seminal work, Grotius asserted that that man existed in all instances in a position of sociability. Even prior to the establishment of official civil institutions, humanity was comprised of communal and coactive social beings. Rather than instinct propelling mankind to cooperate in a united and collaborative effort, individuals were motivated by the realization that society existed in harmony with human nature.[24] Ernst Bloch comments that "Grotius made his starting point . . . (an) optimistic one: It was not the drive for self-preservation motivated by the fear of one's neighbor" as in Hobbes' or Machiavelli's theories of state, "but the social drive, *appetitus socialis*, that led to the social contract."[25] In accordance with Aristotelian and

---

[23] Kelly, 146.

[24] Edwards, 52.

[25] See Ernst Bloch, *Natural Law and Human Dignity* (Cambridge: The MIT Press, 1988), 48.

Stoic tradition, Grotius maintained that from his very evolution, man engaged naturally in civil participation.

Accordingly, in Prolegomena VI which prefaces *De Jure Belli ac Paci*, the Dutch scholar observed that:

> [A]mong the traits characteristic of man is an impelling desire for society, that is, for the social life--not of any and every sort, but peaceful, and organized according to the measure of his intelligence, with those who are of his own kind; this social trend the Stoics called 'sociableness.' (1)

Grotius dismissed the possibility of civil society's existing as an artificial manifestation, and privileged instead humanity's predilection for tranquility and organization gained through communal effort. The naturalness of sociability enabled civilization to benefit man's needs and to bolster his interests. It in no way signaled an annulment of or exodus from nature. By their powers of reason, consequently, human beings were able to comprehend the advantages of social living.

It follows that sociableness served as the natural basis of law, the foundation of a law inherent to humanity. "The maintenance of the social order . . . which is consonant with human intelligence, is the source of law properly so-called" (12), the Dutch scholar stated in Prolegomena VIII. His conviction that sociableness was consonant with human nature established the justification for civil society. In keeping with Stoic belief, he emphasized that municipal society was constructed according to universal human society that had existed in nature, "magna communitas humani generic."[26] Civil society was a microcosm of the larger reality whose purpose was to enable men better to fulfill natural aims. This emphasis had a significant influence on Milton's thought and art. His praise of leadership which promoted common interest and public good are strongly informed by Grotian principles of sociability.

---

[26] Bull, *Hugo Grotius and International Relations*, 72.

Grotius's beliefs are elucidated in the legal tenets he established to represent the interests of the emergent state. H. Lauterpacht's *International Law and Human Rights* contextualizes the Dutch scholar's essential task: "The modern law of nations as it came into being after the Reformation and as it was formulated by Grotius and his immediate predecessors, originated in the historic necessity of regulating the relations of the new sovereign States which arose on the ruins of the temporal unity of Christendom."[27] Grotius confirmed the transformed political milieu of sixteenth and seventeenth century Europe when he indicated that the state "is a complete association of free men, joined together for the enjoyment of rights and for their common interest" (DJBP, I, I, XIV 44). Book 2 of *De Jure Belli ac Paci* defines the sovereign national state which grew out of the loose kingdoms of the Holy Roman Empire:

> An association in which many fathers of families unite into a single people and state gives the greatest right to the corporate body over its members. This in fact is the most perfect society. There is no lawful act of men which does not have relation to this association either of itself or by reason of the circumstances. (II, V, XXIII 253)

A state thereby functioned as a collective assembly of individuals and families affiliated under municipal or civil law. The *ius civile*--formulated in service of regulating this corporate body--though categorized as human law, originated from natural law. Consequently, the civil law and faction of individuals who initiated it were dual manifestations of the natural organization of humanity.[28]

How does warfare fit into this scheme of natural law and sociability? Military strife, it is essential for us to note, existed as a fundamental element in the rise of the state at this time, and was inherent to its very structure. Michael Howard confirms the critical and inevitable role armed combat played in the evolution of the nation-state. The international community was bound together by

---

[27] Hersh Lauterpacht, *International Law and Human Rights* (New York: F.A. Praeger, 1950), 114.

[28] See Edwards, 98.

state protection and formation. "Nationalism," he points out, "[is] almost invariably characterized by militarism . . . could a Nation, in any true sense of the word, really be born without war?"[29] Crucial, additionally, was the worthiness attached to resort to battle. "War was explicitly identified as a positive value, part of a natural process of struggle whereby mankind evolved to ever higher forms of political organization."[30] This veneration of armed strife is incorporated in political theorist Hedley Bull's definition of the state:

> A society of states [or international society] exists when . . .recognizing certain common interests and perhaps some common values, they regard themselves as bound by certain rules in their dealings with one another, such that they should respect one another's claims to independence, that they should honor agreements into which they enter, and that they should be subject to certain limitations in exercising force against one another.[31]

Restricted modes of battle were thus accepted as necessary means to the creation and perpetuation of an orderly and organized society. When contradictions to common interests and values arose, defense of them--sometimes involving war--became essential. The association of martial struggle with fighting for the public good became intrinsic to the maintenance and development of an international society during the Renaissance.

More specifically, war was aligned with the impulse toward sociability which characterized the rise of the nation-state. Under these circumstances, armed strife arose against those who espoused the wrong side of justice: "The following result comes from this principle of sociability: Injustice is that which conflicts with an ordered community of individual rational beings, and all that does not come under the rubric of injustice is just."[32] Consequently, suggests Ernst Bloch, criminal behavior was associated with disordered, irrational

---

[29] Howard, 26-7.

[30] Ibid.

[31] Bull, Kingsbury and Roberts, 7.

[32] Bloch, 49.

impulses. All else which coincided with collective approval fell into the category of condoned behavior. Michael Donelan offers a similarly liberal definition of justice in relation to martial strife. In consideration of motivation for martial activity during Grotius's era, Donelan observes, "war has to be social. It has to be an image for those who believe that they and their states are part of a wider society."[33]

Like the Dutch scholar, whose notion of battle emphasizes the objective of peaceful, good living, but attempts to contend with combat by limiting it in association with law, Milton recognizes the frequency of war in both divine and human realms throughout his work. The poet engages in a complex array of Grotian themes in his military portrayals: the social disposition of man, the moral foundations for participating in rightful battle to protect rights, prohibition against property usurpation. Even in the spiritual and pastoral focuses of his early poetry such as "On the Morning of Christ's Nativity" (1629) and "Lycidas" (1648), the poet alludes to the improbable presence and impact of war.[34]

One of main ways that the poet confirms his adherence to Grotian sociability is in his opposition to despotism. Believing, in keeping with the Dutch scholar, that each individual and the use of his possessions are governed by his own will rather than another's, Milton maintains that there should not be an incompatibility of interests between the individual and state if the state will safeguard personal welfare. In his examination of the relationship between sovereignty and the law in *The Tenure of Kings and Magistrates* and *Eikonoklastes* (1649), for example, he endorses the possibility of resistance against a king who does not benefit the individuals over whom he rules, whose

---

[33] See Michael Donelan, "Grotius and the Image of War, *Millenium* 12 (1983): 233-43, 240.

[34] Images of peaceful, pre-edenic existence evoked in both early works are juxtaposed with allusions to mounting political strife. "On the Morning of Christ's Nativity" alludes to the actual trappings of war chiefly to confirm their absence or distance, and battle imagery emerges primarily in reference to its iconographic significance, in conjunction with stasis. "Lycidas" reveals Milton's increasing objection to corrupt church practices, and suggests revolutionary means to remedy wrong-doing.

monarchical leadership attempts to surpass the legal limits of appropriation in its usurpation of selves and reason. Shortly, we will look at how the paradigm for this abuse of power is embodied in Satan's rulership in *Paradise Lost*, and how the laws of nature, as defined by Grotius, confirm Satan's authoritarianism rather than God's rule.

While analysis of Milton's major early works confirms that war existed as a foremost consideration for him from the start, *Areopagitica* of 1644 most closely anticipates the poet's revolutionary imperative in *Paradise Lost*'s War in Heaven.[35] In this dramatic treatise--similar to the War in Heaven composed twenty years later--Milton depicts martial discord as an essential tool, an occasion to protect and uphold personal liberty as part of an individual's property. Composed during the political upheaval of the early 1640s, the poet's previous references to military strife become concrete in this work, and more discernibly informed by the Dutch scholar's natural law theories. *Areopagitica* served as a reaction to one of the most complete declarations of government censorship in English history, a Parliamentary Act of 1643 which required that all books be licensed by an official censor prior to publication.[36] In his impassioned objection to this prohibition, Milton portrays censorship as an intolerable and antiquated restraint upon the right of individual conscience to decide upon worthwhile reading material for itself. He attempts to prove--by means of an abridged historical account of this mode of suppression--that this formal prohibition was

---

[35] The jurisprudential significance of *Areopagitica* is suggested by the origins of its title, which Milton derived from a famous oration delivered by Isocrates in 355 B.C. Meaning "things to be said before the Areopagus," the name of this work refers to the ancient Greek court of the Areopagus, an illustrious body of three hundred judges that presided on the hill of Ares above Athens. Milton's title suggests a comparison between the Greek institution and the English Parliament, as well as between Isocrates' role in re-establishing Athens' democracy by restoring the court, and his own ("Athens," para. 17-18). Interestingly, Ares--the name of the hill upon which the Areopagus existed--is himself the "God of war," the only son of Zeus and Hera, according to ancient myth.

[36] As James Holly Hanford observes, "This act reflected the increasing determination of the Presbyterian party, now in control in Parliament, to reduce English religious practice and opinion to a new uniformity and to Parliament, to reduce English religious practice and opinion to a new uniformity and to silence political opposition" (*A Milton Handbook* (New York: Appleton-Century-Crofts, 1961), 98).

only appropriate for, and utilized by, degenerate cultures, in particular by the Catholic Church as manifested in the Spanish Inquisition and the Council of Trent.

In his zealous criticism of this interdiction, and in his subsequent demand for freedom of speech--especially in printing--Milton perceives himself immersed in a revolutionary age rife with possibilities for change. Fuelled by the utopian excitement for reformation which prevailed during the early 1640s, *Areopagitica* is discernibly informed by Grotian theories of human liberty and justice in warfare. Similar to *Paradise Lost's* War in Heaven, as we will consider, Milton stresses the value of trial--of enduring tribulation or suffering on the path to truth throughout *Areopagitica*. He prioritizes hard-won truth as the most advantageous and genuine sort. "That which purifies us is triall, and triall is by what is contrary," he informs us near the beginning of this tract (2: 515). "Good books" must be considered alongside "bad books," it follows, for the powers of human reason must be given the opportunity to differentiate between the two. "[K]nowing good by evill" (2: 514)--goodness mixed with or subjected to the vicissitudes and instability of falsity--ultimately produces the most genuine or "solid" good. Consequently, Milton confirms that:

> I cannot praise a fugitive and cloister'd vertue, unexercis'd & unbreath'd, that never sallies out and sees her adversary, but slinks out of the race, where that immortall garland is to be run for, not without dust and heat. (2: 515)

Unable to "praise" a "fugitive and cloister'd" embodiment of Goodness--one who eschews a classically-based "race . . . [for the] immortal garland," and also escapes from a more war-like or earthly confrontation involving "adversary," "dust and heat", the poet stresses that knowledge and survey of vice are imperative to the constitution of human virtue. Freedom of all sorts, confirms Milton, may only be legitimately attained in the course of a sustained struggle with those temptations that are imposed upon the individual from both without and within.[37] In this way, he illustrates how all books serve the cause of truth. Conversely, licensing inhibits the discovery of a new or fuller truth, because it stands in the way of trial and struggle.

This visceral embodiment of necessary trial as a means of attaining Truth is represented by images of martial strife throughout *Areopagitica*. Milton highlights the war-like effort involved in discerning truth from falsehood, virtue from vice. Good and evil must both be encouraged to "sally out," to battle, rather than allow outworn or corrupt practices to remain in possession of the field. Similarly, armed struggle itself must be allowed to be played out, so that truth itself will emerge thus giving justice the opportunity to prevail.

Milton's violent imagery and emphasis upon the war-like characteristics of discourse seem particularly remarkable in reference to the seemingly positive valence he assigns them. Like Grotius, who viewed rightful battle as a chance to defend personal liberty, he metaphorically associates martial strife with the killing or death of antiquated methods, war as the necessary tool for implementing political and social reform. Of particular importance in this context are the numerous passages which suggest the immediacy of the English Civil War as the milieu of discourse. According to Thomas Healy and Jonathan Sawday, in *Areopagitica* "discourse rests upon struggle, almost as though the war being fought in England could be understood as fortuitously available to substantiate Milton's own evocation of knowledge as a species of armed struggle."[38] Francis Barker further points out the work's "remarkable insistence upon the language of killing, blood, and death." "Discourse and violence go, implicitly and explicitly, together in this text, whether one be the object of the other or rather its active

[37] See Andrew Milner, *John Milton and the English Revolution, A Study in the Sociology of Literature* (London: Macmillan Press, 1981), 111.

[38] Thomas Healy and Jonathan Sawday, eds. *Literature and the English Civil War* (Cambridge: Cambridge University Press, 1990), 9.

perpetrator," he explains. "We are more or less forced to encounter in the very language of the text a conception of discourse as struggle or battle."[39]

Furthermore, the poet depicts two disparate female personifications of Truth in *Areopagitica* in association with the tract's efforts to define discourse. Truth, in one version, emerges as a pursued and ultimately dismembered female victim, a "virgin Truth, hewd . . . into a thousand peeces, . . . scatter'd . . . to the four winds" (2: 549). In the other, she appears as a militant warrior: "Let her and Falshood grapple; who ever knew Truth put to the wors, in a free and open encounter" (2: 561), he declares. This diverse representation brings to mind the "so many cunning resemblances" (2: 514) that render it difficult to differentiate between truth and falsehood and make truth hard to recognize. Truth's doubled portrayal, furthermore, challenges the reader to determine which definition--perhaps both--encompasses the "truest" metaphor. Milton suggests his faith in the power of his reader's rational consciousness to solve this question for itself.

This double representation of Truth--in keeping with the optimistic nature of Milton's scholarship throughout this period--exhibits what Andrew Milner designates Milton's "equality of opportunity"[40] model of human behavior in reference to man's occasion to choose between good or evil on his own. In this way, the revolutionary imperative contained within these works is directly associated with an opportunity for freedom and lack of external restraint. Toward the beginning of *Areopagitica*, the poet observes

> To the pure all things are pure, not only meats and drinks, but all kinde of knowledge whether of good or evill; the knowledge cannot defile, nor consequently the books, if the will and conscience be not defil'd. For books are as meats and viands are; some of good, some of evill substance.
> (2: 512)

He confirms the morally upstanding or "pure" man's inherent incorruptibility by evil. More important than man's proximity to and engagement with "good books"

---

[39] Ibid., 96.

[40] Milner, 110.

--which censorship would arrange for by coercive means--is his actively engaging in literature as an enactment of the inherent "goodness" of his rational mind. Later in the tract, in reference to this emphasis upon rational choice, Milton presents inner good and undefiled will of conscience with his example of Adam:

> [W]hen God gave [Adam] reason, he gave him freedom to choose, for reason is but choosing; he had bin else a meer artificiall Adam, such an Adam as he is in the motions. We our selves esteem not of that obedience, or love, or gift, which is of force: God therefore left him free, set before him a provoking object, ever almost in his eyes; herein consisted his merit, herein the right of his reward, the praise of his abstinence. (2: 527)

An "artificial" Adam as he is "in the motions," a marionette manipulated from without, is far less admirable, the poet suggests to us, than one who chooses his own movement. Portraying God as a creator who directly endows man with the powers of reason or "freedom to choose," Milton confirms the source of Adam's free will.

Not merely, we can argue, does Milton rally against the efficacy of external constraint--the imposition of a factitious peace of forced conformity--in this tract. More notably, he demonstrates that this hindrance upon freedom cannot effectively take hold. Knowing good by evil, the arduous "trial" of choosing good with evil in seductive proximity, is the only true path. In Milton's system of morality--as in Grotius's-- this necessary though difficult requirement supersedes the efficacy of coerced conformity. The poet's confidence in his fellow Englishmen's ability to identify good by the powers of their rational choosing makes this so. According to James Holly Hanford, in *Areopagitica* "[the poet] deals with the impossibility of the attempt to make men virtuous by external constraint. Corrupting influences are present everywhere and can be met only by building up an inner discipline and the power of rational choosing."[41] Andrew Milner says essential the same: "Milton believed that all men are essentially free

---

[41] Hanford, 100.

by nature, and that they cannot therefore properly be subject to the tyranny of external constraint".[42]

Strikingly, Milton's emphasis upon necessary trial and the war-like effort involved in discerning truth from falsehood in *Areopagitica* is sustained and further clarified in *Paradise Lost*. As the poet completed his epic, his hopes that the English Revolution would produce a permanent republican state were disappointed when England was re-"monarchized" (*The Ready and Easy Way to Establish a Free Commonwealth* (1660), 7: 462) in 1660 under King Charles II. In *Paradise Lost*, it seems to me, Milton adheres to Hugo Grotius's version of natural law to differentiate between evil and good at a time when evil seemed to prevail. He strives to make the laws of nature perform beyond the restoration of monarchical rule to confirm God's just purpose in permitting the unfavorable outcome of the English Civil War to occur. Confirming the poet's Grotian belief that the parameters for legitimate military action must be circumscribed in accordance with the laws of nature, two types of armed conflict--illegitimate warfare originated by Satan, and God's justified defensive battle in reaction--take place in the War in Heaven in Book 6 of the epic.

This War, it is significant to note, is inspired by Satan's angry reaction to the Son's promotion--"him hav[ing been] anointed" (5.605) and "[a]ll power on him transferr'd" (6.678)[43] by God. At the start of the epic, we encounter Satan just after he and his troops have been hurled down "[w]ith hideous ruin and combustion down/To the bottomless perdition" (1.46-7) of Hell. Despite this set-back, Satan portrays himself as a defender of the law who seeks to inspire his legions to continue to wage their attack against a criminally-intended God. According to Satan, the Son's advancement represents a perversion of natural justice. As James A. Freeman remarks, "God has not simply stolen Satan's 'just

---

[42] Milner, 109.

[43] The poetry of John Milton is quoted from *The Poems of John Milton*. Ed. John Carey and Alastair Fowler (Longman: London and New York, 1968), and is cited within the text.

right'; he has also fitched from Satan's followers their 'just inheritance.'"[44] In a bold abuse of power, claims Satan, God acts as a usurper. Father and Son undertake to "extend / His empire" (2.315, 325-7), resolving to acquire territory which exists beyond their kingdom in heaven. Now God's rule will extend throughout Hell, where his authority will be "arbitrary" (2. 334). In reaction to this corrupt vie for power, Satan sanctions the right to wreak vengeance against the act of territorial misappropriation which he alleges God is committing.

In a skewed representation of the revolt Milton himself endorsed years before in the *Tenure of Kings and Magistrates* (1649) and *Eikonoklastes* (1649), it seems to me, Satan attempts to convince his regiments how to overthrow a ruler who is indifferent to legality. Becoming even more hardened, "more haughty" (5.852) and "fearless" (875), he moves "with vast . . . strides" (6. 109) into battle. Satan looks to combat for the solution to what he perceives as God's abuse of power. His false claim at being wronged by God results in his initiation of unlawful battle. Satan's criminal aspirations, significantly, are centered around territorial usurpation, a primary Grotian category of illegitimate battle. "[Unjust] cause . . . [includes] the desire for richer land" (DJBP, II, XXII, VIII 550) which is the possession of another, the Dutch scholar informs us. Challenging the laws of property ownership, those which encompass the first and most essential element of justice, "The banded powers of Satan hasting on / With furious expedition; for they weened / That self same day by fight, or by surprise / To win the mount of God and on his throne / To set the envier of his state, the proud / Aspirer" (6. 85-90). In this intimate juxtaposition of good and evil--in which Satan commits the very crime he wrongfully accuses God of committing--Milton demonstrates, as he did in *Areopagitica*, how difficult it is to distinguish between truth and falseness. He seeks to show us, furthermore, that they reside in irksome proximity as two kinds of signs leading to the creation and confirmation of genuine goodness.

---

[44] James A. Freeman, *Milton and the Martial Muse* (Princeton: Princeton University Press, 1980), 146.

Crucially, Milton chooses to designate the nucleus of his poem as an interlude of martial strife--a battle conducted by the rebel angels against the good. In Book II of *De Jure Belli ac Pacis*, Grotius urges that "war is not to be undertaken save from the most weighty cause at the most opportune time" (II, XXIV, IX 575). With its intricately-constructed earthly trappings of martial strife, the authenticity of Milton's imagery confirms that actual war proceeds here as a sanction to right wrongs. Taking place over only three days time, this abbreviated battle confirms Milton's desire to temper war's portrayal. Occurring as a necessary instrument, and conducted within rightful limits, the War in Heaven is modeled on the Dutch scholar's concept of martial conflict. Choreographed in the service of showing some greater point of supreme necessity, this interlude underscores Grotius's opposition to the brutish methods of battle which characterized war during his own time. Circumscribed in accordance with the laws of nature, as an opportunity to protect and maintain personal liberty, military action need not be irreconcilable with justice.

By the conclusion of the first day of this struggle, God's troops prevail but are unable to expel Satan and his cohorts--though "with pale fear surpris'd" (393)--out of Heaven. On Day Two, Satan's ambition to commit far-reaching devastation, to "invent / With what more forcible [he] may offend" (464-5), is dramatized by his fabrication of the cannon and formulation and first use of gunpowder. As a derivative of his all-consuming aspiration to destroy, the Devil's tool of territorial annihilation may be perceived as an attempt to counteract the effect of his dwindling stature within the poem. Longing for "more valid arms" (6.438), the rebel troops desecrate heavenly territory in order to obtain gunpowder, and then load their tool and fire it:

> . . . . up the turn'd
> Wide the celestial soil, and saw beneath
> The originals of nature in their crude
> Conception; sulphurous and nitrous foam
> They found, they mingled, and with subtle art,
> Concocted and adusted they reduc'd

To blackest grain. (6. 509-15)

Defiling the land for their unlawful purposes, Satan's forces eagerly exploit nature in service to their criminal war operation. Explicitly contradicting Grotian requirements for just war, the Devil's mechanical contrivances ravage the divine terrain.

Nonetheless, Milton presents this show of technological innovation in order to thoroughly study Satan's illicit intentions --God from his throne in Heaven had "foreseen / This tumult, and permitted all" (6. 673-74), thus allowing this act of destruction to be completed in its entirety. Providing this naturalistic representation of the Devil's creation of cannon and manufacture and use of gunpowder, the poet facilitates the development of his readers' observational skills--he insists on the importance of their witnessing this mechanical contrivance, so to speak, with their own eyes.[45] Milton's focus on the production and implementation of this chemical technology, in and of itself, reflects enthusiastic attitudes toward scientific advancement which characterized the Renaissance. It supports the popular notion that innovation in weapons of warfare could aid in the perpetuation of an orderly and organized society, and help man attain higher degrees of political organization. And yet, placing gunpowder and cannon in the hands of Satan, Milton more specifically demonstrates what can happen when the fruits of this martial technology are deployed by evil. He writes for an audience for whom war was becoming increasingly destructive, and was only in the process of being codified according to law. His reader's ability to differentiate between good and evil had become more imperative than ever before.

[45] Investigating the emergence of the new mechanistic outlook which characterized the Renaissance (1540-1687), James R. Jacob confirms that at time there "was the idea that science should be less about the contemplation and demonstration of old truths than about the discovery of new ones, even if this meant manual operations, going directly to nature, and getting one's hands dirty in the bargain" (*The Scientific Revolution: Aspirations and Achievements, 1500--1700* (Atlantic Highlands, NJ: Humanities Press, 1998), 53).

Accordingly, Milton counterbalances his demonstration of Satan's criminal efforts at appropriation and possession with God's permissible defensive battle. As a response to Satan's criminal efforts at appropriation and possession, God's troops answer with permissible defensive battle. "No other just cause for undertaking war can there be except injury received" (DJBP, II, I, IV 170), the Dutch scholar confirms in *De Jure Belli ac Paci.* In the second book of this treatise, which he devotes to identifying just causations for battle, he explains that "Authorities generally assign to wars three justifiable causes, defense, recovery of property, and punishment. All three you may find in Camillus's declaration with reference to the Gauls: 'All things which it is right to defend, to recover, and to avenge'" (II, I, IV 171). In keeping with these parameters, the good angels respond to the bombardment of cannonballs by hurling mountains at their attackers

> Light as the lightning glimpse they ran, they flew,
> From their foundations loosening to and fro
> They plucked the seated hills with all their load,
> Rocks, waters, woods, and by the shaggy tops
> Up lifting bore them in their hands (6. 642-46)

As "power / Which God hath in his mighty angels placed" (6. 637-38), these mountains serve as the "ultimate natural weapon" (Bennett 89)--organic munitions which serve as a powerful means of counteraction. In reaction to this unforeseen tactic, "terror seized the rebel host, / When coming towards them so dread they saw / The bottom of the mountains upward turned" (6. 649).

At the beginning of *De Jure Belli ac Paci,* Grotius further demonstrates that self-preservation is the first and most fundamental principle of the natural law, employed for positive ends in order to safeguard rights. In necessary self-protection such as this, "against such hellish mischief fit to oppose" (6. 636), force is permissible to preserve legitimate rights, and therefore does not contradict the law. Pursued in service of the laws of nature, military strife confirms that justice is existent and operative. In keeping with this necessity, the task of identifying and safeguarding ownership are primary considerations for Milton and

Grotius. Under the laws of nature, the Dutch scholar tells us, war is shaped by the powers of man's reason and his innately social nature. Ethical motives inspire individuals to engage in legitimate martial combat--to defend their rights. Accordingly, the military measures pursued by the loyal angels represent and defend the laws of God. Their response takes place in a Grotian manner as an essential instrument--in order to do away with criminal war, one must pursue just war.

Even so, the War in Heaven persists. Despite the counter-attack by God's troops:

> . . . . horrid confusion heaped
> Upon confusion rose: and now all heaven
> Had gone to wrack, with ruin overspread (6. 668-70)

The divine landscape is progressively ravaged. In reaction to this widespread chaos, the arrival of the Son on Day Three achieves a genuine restitution of Heaven. Saving the culmination of the apocalyptic spectacle of celestial battle for the Son, God informs him "that the glory may be thine / Of ending this great war, since none but thou / Can end it" (701-04). In a supreme act of devotion, the Son is the only one who can deliver the angels from never-ending warfare. Descending at the end of Book VI, the conquering Son in a triumphal chariot materializes in a commotion of flames and smoke:

> Ascended, at his right hand Victory
> Sat Eagle-wing'd, beside him hung his Bow
> And Quiver with three-bolted Thunder stor'd,
> And from about him fierce Effusion roll'd
> Of smoke and bickering flame, and sparkles dire;
> Attended with ten thousand thousand Saints,
> He onward came, far off his coming shone (6. 762-8)

Arriving miraculously in a self-driven carriage, "the chariot of paternal deity/Flashing thick flames" (6. 750-1), his effortless engagement in the War contrasts with his evil adversaries' painstakingly-constructed cannon and gunpowder. The poet accentuates the ease with which the Son enters and resolves this conflict. He attains victory without trial--proceeds "instinct with Spirit" (752)--due to his supreme dedication to God. The Son's recovery of order appears as a supreme act of justice within the poem--the culmination of defensive warfare in which the laws of God are confirmed.

Portrayed in conjunction with Grotius's specifications for rightful battle, furthermore, the notion of limited warfare is essential to the Son's intervention. The Son commends his troops for their zealous efforts--"faithful hath been your warfare, . . . fearless in his righteous cause" (6. 803-04), but emphasizes that Satan's fury is ultimately pointed toward him. Having delivered this reminder, he subsequently engages in an impressive show of prowess, "in his right hand / Grasping ten thousand thunders, which he sent / Before him" (835-37), impelling the rebel forces to drop their weapons in fear. Notably, the Son's display of force is juxtaposed with a call to *temperamenta.* Despite his show of strength, Milton emphasized, the Son tempers his martial measures: "Half his strength he put not forth, but checked / His thunder in mid volley, for he meant / Not to destroy, but root them out of heaven" (6. 853-55). Consequently, he avoids excessive violence. "Devastation should be refrained from," the Dutch scholar tells us, "if there is good hope for a speedy recovery" (DJBP, III, XII, III 748). With the Grotian concept of just battle foundational to the War in Heaven, it seems, the timely achievement of peace is prioritized.

The settlement the Son facilitates on the third day of this battle harks back, it seems, to Milton's earlier "On the Morning of Christ's Nativity," where the coming of the Son is similarly affiliated with peace and tranquility guarded by militarily-clad angels. Taking on the task of dealing with Satan and his cohorts, he counsels the faithful angels:

> Stand still in bright array ye saints, here stand
> Ye angels armed, this day from battle rest;
> Faithful hath been your warfare, and of God
> Accepted, fearless in his righteous cause,
> And as ye have received, so have ye done
> Invincibly; but of this cursed crew
> The punishment to other hand belongs (801-07)

Similar to the tranquility associated with the arrival of the Son in "The Nativity Ode," this representation suggests a time when war shall end.[46] Altering the "perverse commotion" (706) of martial discord into a realm of peace, Milton ends the War in Heaven--like his "Nativity Ode"--by associating the achievement of Justice with the attainment of serenity.

Depicting the War in Heaven in association with Grotian theory, Milton differentiates between the rebel troops' imposition of criminal battle and the good angels' legitimate and necessary struggle for justice. Defeated in the conflict they began--expelled out of Heaven and forced to fall flaming for nine days though Chaos until they arrive at Hell--the demise of Satan's forces confirms that their initiation of illegitimate warfare gave them no superiority over God. Instead of serving as an occasion for Satan to display his martial skill, this conflict further demeans him. The technocrat of the cannon and gunpowder temporarily desecrates Heaven, ravishing the natural order of divine territory, but does not come close to overtaking the authority of God. By contrast, the Son's moderated effort serves to successfully expunge his enemies. And prior to this, the faithful angel's employment of natural arms--propelling mountains at the onslaught of cannon-fire, a legitimate target of defense--proves to be a much more justified act of self-preservation.

Providing his readers with an intimate look at Satan's fabrication and first use of cannon and gunpowder, Milton exposes them to the subtle nuances of innovative martial weaponry used for unlawful purposes. The poet enables his readers to "know good by evill" (*Areopagitica*, 2: 514) with the aim of achieving a more genuine understanding of truth, underscoring his belief that knowledge and survey of vice are imperative to the constitution of human virtue. He places God's and Satan's forces in close proximity as they battle it out across the divine terrain, and has Satan accuse God of committing the criminal act of territorial

---

[46] In both poems, Milton seeks to emphasize that the Son's war-like participation derives its significance from the spiritual rather than physical realm, a subject which he closely investigates in *Paradise Regained.*

usurpation he himself commits. In this way, Milton challenges his audience to distinguish between God's legitimate efforts and Satan's ill-intended ones. Focusing on the micro-movements of the illegitimate use of this developing chemical technology, the poet encourages his readers to make an informed choice, in favor of martial strife pursued in the service of justice and freedom.

Elizabeth Oldman (Tulane University) is Visiting Assistant Professor in Early Modern British Literature at Tulane University. She received her Ph.D. in English from New York University and has also studied at Leiden University and The Hague. Her research interests include 17th-century British Poetry and Prose, the History of Law and Political Thought, and the Literature and Philosophy of War. Her most recent essays--both about John Milton and 17th century warfare--are forthcoming in *Studies in Philology* and *War, Literature, and the Arts*.

To Mine for Truth:
The Metaphor of Mining in Francis Bacon's *The Great Instauration*

Lauren Klein

Francis Bacon (1561-1626) intended *The Great Instauration* to revolutionize science. In his *magnum opus*, Bacon outlines and subsequently models a new method for scientific discovery, one that emphasizes observation and experience over contemplation and theory. Bacon's writing is deliberate and precise, with language as meticulously crafted as the logic it espouses. In his examination of the scientific practices of the time, he employs several powerful metaphors. Bacon despairs that men fail to "explore beyond" what is accessible and apparent (*N*O: 7).[1] He laments the propensity of even great scientists to attribute the limitations of their theories to the difficulty of probing for the "hidden recesses of truth" (*NO*: 10). Bacon believes in overcoming this complacency through a process of exploration. He envisions an enlightened world filled with knowledge and truth, but a truth that requires effort to be attained.[2]

In the first lines of "The Plan of the Work," Bacon outlines the process of "a form of induction... which is drawn not only from the inner recesses of the mind, but also from the very bowels of Nature" (*NO*: 21). By analogizing the brain to the "bowels of Nature," Bacon implies a connection between the process

---

[1] References to the works of Francis Bacon are abbreviated as follows: *AL: The Advancement of Learning,* ed. Michael Kiernan (New York: Oxford UP, 2000); *NO: New Organon*, trans. Peter Urbach and John Gibson (Chicago: Open Court, 2000); *SS: Sylva Sylvarum,* trans. Basil Montagu (London: William Pickering, 1826); *NA: New Atlantis* (Kila: Kessinger).

[2] In addition to the preface and "The Plan of the Work," Bacon completed only two sections of the proposed six-part project: *The Advancement of Learning* (1605) and *New Organon* (1620). Most scholars consider *Sylva Sylvarum*, published posthumously in 1667, to comprise part of the third section Bacon describes in his outline as *The Phenomena of the Universe*. Given the immense scope of the project, it is not surprising that it remained unfinished at the time of his death. Nevertheless, the extant sections stand as a monumental treatise on natural philosophy and the scientific method.

of mental exertion and the physical exertion required to plumb the depths of the earth. In doing so, Bacon proposes a new theoretical foundation for science, one that endeavors to "sink the beginnings deeper than men have ever done" (*NO*: 21). He urges men to dig down to truth, to "search and mine into that which is not revealed" (*AL*: 184). In his appeal to seek out the truth of nature, Bacon repeatedly employs the image of the underground. By insisting on a literal excavation of knowledge, Bacon accelerates the exploration of scientific truths, as well as the exploration of the underground itself.

The twentieth-century scientist Stephen Jay Gould, who resembles Bacon in his emphasis on the "big picture," that is, on the implications of his observations and experiments on science as a whole, states that "the interplay of internal and external sources—of theory informed by metaphor and observation constrained by theory—marks any major movement in science."[3] Indeed, the publication of *The Great Instauration* marks a major scientific event. Bacon's stance is a radical one, rejecting centuries of emphasis on contemplation in place of a new, activist theory of discovery. His language is nothing if not intentional. His repeated exhortations to dig for knowledge and mine for truth are incantation-like. By invoking images of mining, Bacon gains a valuable, if complex metaphor, one that both informs and is informed by the theories he espouses.

The history of the mine is bound up in complex issues of nature, gender, and religion. Carolyn Merchant, in *The Death of Nature*, traces the origin of this complexity to the earliest conception of the earth: the belief that the earth was alive and, specifically, female. According to Merchant, this organic conception of the earth took two forms:

> Central to the organic theory was the identification of nature, especially the earth, with a nurturing mother: a kindly beneficent

---

[3] Stephen Jay Gould, *Time's Arrow, Time's Cycle: Myth and Metaphor in the Discovery of Geological Time* (Cambridge: Harvard University Press, 1987), 8.

> female who provided for the needs of mankind in an ordered, planned universe. But another opposing image of nature as female was also prevalent: wild and uncontrollable nature that could render violence, storms, droughts, and general chaos. Both were identified with the female sex and were projections of human perceptions onto the external world.[4]

Although these two views of a female earth—one as maternal and beneficent, the other as wily and deceptive—seem at odds with each other, they converge in their adherence to the personification of nature as a woman.

When studied within the context of mining, the conception of the earth acquires additional female attributes. Aristotle, and other early thinkers, believed that metals owed their origin to celestial rays and vapors, which in turn "g[a]ve birth" to metals inside the earth.[5] Other philosophers inferred that metals grew like trees, with "mineral veins" that resembled "branches and twigs," or like "seeds" that matured in underground "matrices" or "wombs."[6] Merchant summarizes, "For most traditional cultures, minerals and metals ripened in the uterus of the Earth Mother, mines were compared to her vagina, and metallurgy was the human hastening of the birth of the living metal in the artificial womb of the furnace."[7] The view of the earth as female and her womb as the source of metals imposed what Merchant terms a "normative constraint" on the practice of mining. She comments: "As long as the earth was considered to be alive and sensitive, it could be considered a breach of human ethical behavior to carry out destructive acts against it."[8] Rosalind Williams, in *Notes on the Underground*, echoes Merchant's characterization of mining as "an enterprise of dubious

---

[4] Carolyn Merchant, *The Death of Nature: Women, Ecology, and the Scientific Revolution* (San Francisco: Harper and Row, 1980), 2.

[5] Johan Glauber, quoted in Frank Dawson Adams, *The Birth and Development of the Geological Sciences* (New York: Dover, 1954), 281.

[6] Ibid., 286, 290.

[7] Merchant, 4.

[8] Ibid., 3.

morality, comparable to mutilation and violation."[9] As a result of the overarching conception of the earth as female, the act of mining was conceived of as a rape of nature.

The symbolism found in the writing of the ancients reflects the early, negative connotations of mining. In the *Natural History*, the Roman scholar Pliny the Elder (AD 23-79) rails against what he perceives as the intrusive conquest of Mother Nature: "We penetrate into her entrails, and seek for treasures... as though each spot we tread upon were not sufficiently bounteous and fertile for us."[10] Pliny's depiction of mining as a vivisection of the earth suggests the reason for his condemnation of the practice, and his condemnation extends to those who participate in it. Miners, he believes, are disrespectful of the earth and motivated by greed. In the *Metamorphoses*, the poet Ovid (43 BC-AD 17) builds upon Pliny's characterization of the miner as an avaricious ingrate. He writes:

> The rich earth
> Was asked for more; they dug into her vitals,
> Pried out the wealth a kinder lord had hidden
> In stygian shadow, all that precious metal,
> The root of evil. They found the guilt of iron,
> And gold, more guilty still. And war came forth.[11]

Ovid underscores Pliny's representation of mining as a destructive act against mother earth, and conflates it with a more virulent sort of greed. To Ovid, the mine's repugnant contents, wrested from the "stygian shadows" of the underworld, lead directly to war. His perspective is acute; the pursuit of gold and iron provide the impetus for war as well as the materials for battle. Lewis

---

[9] Rosalind Williams, *Notes on the Underground* (Cambridge: Massachusetts Institute of Technology Press, 1990), 24.

[10] Quoted in Merchant, 30.

[11] Georgius Agricola, *De Re Metallica,* trans. Herbert Clark Hoover and Lou Henry Hoover (New York: Dover, 1950), 7.

Mumford, in his landmark text, *Technics and Civilization*, employs the link forged by Ovid to support his argument about the influence of mining on the development of modern warfare. According to Mumford, war and mining "played into each other's hands."[12] This association may have provided an additional normative constraint that curtailed the practice of mining and limited the inquiry into its origins and development.

The impact of the Christian value system strengthened the cultural sanctions against mining. By the early sixteenth century, the conceptualization of mining as a rape of the earth carried with it the additional burden of sin. The avarice that the ancients first associated with the mine was reframed in terms of God. The earth was His creation, and the search for what God had so intently hidden from sight implied ingratitude and greed. Edmund Spenser (1552-1599) demonstrates how Christian morality was imposed on the cultural construct of the mine. In the *Faerie Queene*, Sir Guyon visits the Cave of Mammon, where he observes: "Then gan a cursed hand the quiet wombe / Of his great Grandmother with steele to wound, / And the hid treasures in her sacred tombe / With sacrilege to dig."[13] Spenser reiterates the themes of antiquity, but adds a religious connotation. Mammon's digging is a "sacrilege" performed by "cursed" hands. Not only does Mammon defile the warm, matronly earth, but he also desecrates it. In *Paradise Lost*, John Milton (1608-1674) presents a similarly sacrilegious characterization of mining. His Mammon leads men into the earth where, by his instruction, they "with impious hands /Rifl'd the bowels of thir mother Earth / For treasures better hid."[14] In characterizing visitors to the underground as greedy, "impious" thieves, Milton, too, suggests a lack of religious conviction: that is, man should not seek out that which God has so clearly intended to keep out of

---

[12] Lewis Mumford, *Technics and Civilization* (New York: Harcourt, Brace & World, 1962), 76.

[13] Book II, Canto VII, ll. 145-8, quoted in Merchant, 38-9.

[14] Book I, ll. 685-90, quoted in Merchant, 39.

human hands. Both Spenser's and Milton's depictions of Mammon illustrate the dominant cultural and religious views that curtailed the development of mining and other related scientific discoveries.[15]

Practical considerations also limited underground exploration. Because, in the sixteenth and seventeenth centuries, coal could still be found in abundance in open pits, there was little incentive to analyze its origins. Roy Porter, in *The Making of Geology*, explains, "It was widely assumed—not the least by working miners—that strata were completely random and jumbled, and the terrain utterly disrupted by faults, caverns, [and] dykes."[16] Furthermore, the environment of the mine was especially dangerous. Both the frequency of explosions and the poor ventilation contributed to a high fatality rate for miners and others who ventured down into the earth. In his history of British coal mining, Gilbert Stone describes early mining pits as "coal seam[s] by which geological action has been thrown upwards or has been disclosed so that the coal breaks out from the surface of the earth."[17] Even within the context of his factual account, Stone conjures up vivid images of the mine. He evokes a picture of a violent underground consistent with the perception (and reality) of mining as a dark and dangerous practice.

---

[15] John Donne (1572-1631) takes up these issues in his love poetry, but in a different manner from Spenser and Milton. Unlike these poets, who personify the mine as female, Donne describes women in terms of the mine. In "An Anatomy of the World," he speaks of woman's "refined" virtue and the "alloys" of her mind (275 ll. 175-176). "Love's Alchemy" begins with similar symbolism: "Some that have deeper digged love's mine than I, / Say ,where his centric happiness doth lie" (65 ll. 1-2). This reversal allows a more sensual, beneficent conception of mining to emerge. Donne's mine is slightly dark and illicit, but no longer associated with the violence and sacrilege that derive from the obverse image. See also "Love's Progress" for a longer passage with revealing topography: two kinds of love, a virtuous one above the earth and an underground Cupid that lives in evocative "pits and holes" (122 ll. 16, 27-36). Donne's progressive vision reflects the changing perceptions of the mine of the early seventeenth century that, I will later argue, Bacon's own work sets in motion. John Donne, *The Complete English Poems*, ed. A.J. Smith (New York: Penguin, 1996).

[16] Roy Porter, *The Making of Geology: Earth Science in Britain 1660-1815* (New York: Cambridge UP, 1977), 58.

[17] Gilbert Stone, *The British Coal Industry* (London: J. M. Dent & Sons, 1919), 4.

Critics note that Bacon is "unaccustomedly quiet" on the subject of metallurgy and the mine.[18] Bacon's few observations on mining, embedded in the final pages of *Sylva Sylvarum*, discuss the effects of certain vapors on the human body, and describe the dangers of the practice: "It happeneth often in mines, that there are damps which kill, either by suffocation, or by the poisonous nature of the mineral: and those that deal much in refining, or other works about metals and minerals, have their brains hurt and stupefied by the metalline vapours" (*SS*: 497). It is noteworthy that Bacon chooses to comment on the brain-dulling effect of the metallic vapors, since the scientific merits of mining were still being contested.

In fact, Georgius Agricola (1494-1555), whom many consider to be the founder of modern metallurgical science, begins his treatise on mining, *De Re Metallica*, with an extended justification of mining as an art. He states, "Many persons hold the opinion that the metal industries are fortuitous and that the occupation is one of sordid toil, and altogether a kind of business requiring not so much skill as labor. But as for myself, when I reflect carefully upon its special points one by one, it appears to be far otherwise."[19] In addition to the normative constraints, the religious prohibitions, and the intrinsic danger of the mine, a major source of the resistance to mining as a scientific pursuit was the low social status of the miners themselves. Prior to the seventeenth century, the only people who journeyed below the surface of the earth were slaves, criminals, and prisoners of war. Mumford explains, "Mining was not regarded as a humane art: it was a form of punishment: it combined the terrors of the dungeon with the physical exacerbation of the galley."[20] Seventeenth-century England was still a

---

[18] Porter, 16.

[19] Agricola, 1.

[20] Mumford, 67. Interestingly, Mumford also recounts how women were employed to work in the cramped passages of the mine. He explains, "Until modern tools were invented, the passage itself was a cramped one: to extract ore, children and women were employed from the earliest days to crawl along the narrow tunnel, dragging a laden cart: women indeed were used as beasts of burden in English mines right up to the middle of the nineteenth century" (68). This is a

highly stratified society. There was little reason to elevate the drudge work of servants and felons into a practice worthy of intellectual study.

For these reasons, it is credible, as Porter asserts, that investigation of the underground "scarcely advanced in Britain in the first half of the seventeenth century."[21] But if innovation in mining and its related techniques lagged so far behind developments in other technological fields, why did Bacon seize upon the mine as a major structural metaphor of his theory? Perhaps his view can be ascribed to what Hannah Arendt, in *The Human Condition*, describes as the "pathos of novelty" that characterized much of the writing of the seventeenth-century natural philosophers.[22] Or, possibly, Bacon perceived a connection between the literal process of extracting metals from the earth and his proposed technique for the acquisition of knowledge. Did the issues that men encountered in their attempts to advance mining technology parallel those that Bacon confronted in his endeavor to advance scientific learning? Did the mine provide a real world model for Bacon's theoretical method of scientific discovery? Perhaps Bacon identified the resonance between his idea of the mine as an environment subjected to control by man and his view of nature as similarly controlled. Was it the prospect of a markedly technological future that appealed to Bacon? Did the mine provide him with a cultural construct that enabled him to rework the past? Or could the forbidden recesses of the mine provide Bacon with a safe haven for a potentially dangerous scientific exploration? By situating his work in the already forbidden world of the underground, did Bacon believe that he could protect himself from accusations of heresy? These questions penetrate to the core of Bacon's preoccupation with the provocative image of the mine.

---

fascinating irony: women being exploited in the process of participating in the exploitation of the earth.

[21] Porter, 16.

[22] Hannah Arendt, *The Human Condition* (Garden City: Doubleday-Anchor, 1959), 226.

As a rhetorical device, the metaphor of the mine strengthens Arendt's claim of the "pathos of novelty." To be sure, Bacon intends *The Great Instauration* as an intellectual call to arms. He indulges in what Arendt identifies as "the almost violent insistence of nearly all the great authors, scientists, and philosophers since the seventeenth century that they saw things never seen before, thought thoughts never thought before."[23] Bacon's self-description of his project, to "sink the beginnings deeper than men have ever done," fuels Arendt's claim of sensationalism (*NO*: 21). Bacon employs the metaphor of the mine to emphasize the uncharted territory of his new philosophy. The cultural context of the mine lends an element of the illicit, an attribute not often associated with scientific inquiry. At a time charged by the emergence of competing scientific theories, it is quite plausible that Bacon might have recognized in mining not only a metaphor for positioning his project, but also a method for introducing additional intrigue, and hence increased interest, in his scientific study.

By describing *The Great Instauration* in terms of exploration and discovery, Bacon also aligns his work with the British colonial enterprise. In *New Science, New World*, Denise Albanese connects the projects of science and colonialism in terms of their shared emphasis on novelty. She explains, "Expeditions to America and scientific programs both propagandize themselves as voyages out into uncharted territory, where the sense of excitement that attaches to new ventures covers over the work of domination that underwrites exploration of the globe and of nature."[24] By linking the process of exploration to practices of concealment and "domination," Albanese invokes, if unintentionally, the violent rhetoric of the mine. Bacon's claim, in "The Plan of the Work," that he will "root out and drag from the darkness many things that no one would think of investigating," illustrates these processes (*NO*: 26). Bacon's language is

---

[23] Ibid.

[24] Denise Albanese. *New Science, New World* (Durham: Duke University Press, 1996), 97.

simultaneously constructive and destructive. He emphasizes the originality of his ideas, but at the expense of nature. By describing a process of "rooting out," Bacon suggests that his new method will disrupt the natural order of things. He conjures up an image of an organic system, one in which elements are bound together by a network of roots, then literally uprooted, and replaced by an orderly, methodical, scientific system.

Bacon also employs the concept of mining in a pragmatic sense, to exemplify the application of his new scientific method. It is well documented that Bacon read Agricola's writing on metallurgy.[25] Quite possibly, it was this text that enabled Bacon to identify the parallels between the project of mining and his own endeavors. Agricola addresses many of the failings of science that Bacon endeavors to correct—the tendency to emphasize theory over observation, as well as what Bacon describes as the "almost spell-bound reverence for antiquity" (*NO*: 92). A large portion of Agricola's work deals with refuting misconceptions that emerge from ancient speculation. In *De Ortu et Causis Subterraneorum*, Agricola relates, with some indignation:

> Aristotle maintained that the metals are formed from watery vapors. But why does he not submit some basis of proof for this assertion? And likewise he asserts that the infusible rocks are developed out of dry exhalations. Why does he assert this without any evidence? The very mountains themselves refute such a statement.[26]

Perhaps Bacon's insistence on the primacy of observation derives in part from Agricola's frustration that scholars repeatedly fail to observe and experiment before they theorize. When Agricola undertakes the project of composing *De Re Metallica*, he vows to adhere solely to confirmed evidence: "I have omitted all those things which I have not myself seen, or have not read or heard or from

---

[25] Paolo Rossi, *Francis Bacon: From Magic to Science,* trans. Sacha Rabinovitch (Chicago: University of Chicago Press, 1968), 8.

[26] Quoted in Adams, 308.

persons upon whom I can rely."[27] This methodology is almost identical to that articulated by the natural philosophers, although Agricola predates them by decades. Years before Bacon proposed his "form of induction that will analyze experience and take it apart," miners were already learning from experience (*NO*: 21). The constraints of working in the mine accelerated related scientific practices, and provided Bacon with a practical setting for the development of a theory informed by observation.

In describing the elements of his scientific method, Bacon draws not only on lessons learned from the history of mining but also on aspects of the process of mining itself. When Bacon formulates his response to centuries of compounded misinformation, he describes his solution in terms of mining and machinery. He explains, "The whole operation of the mind must be completely re-started, so that from the very beginning it is not left to itself, but is always subject to rule; and the thing accomplished as if by machinery" (*NO*: 38). Calling for new rules and methods, Bacon proposes a "way of discovering science" that is quite similar to the means by which miners "assay and prove" the ore of the earth.[28]

Bacon's exhortation to operate the mind, like the mine, "as if by machinery," implies not only a mechanical method of observation but also a sophisticated model for contemplation and self-critique. The central premise of *Humanism, Machinery, and Renaissance Literature*, by Jessica Wolfe, is that the machine functions in its primary capacity as a mediator:

> Machinery mediates some of the most urgent political and intellectual problems vexing humanistic culture. Yet machines are also mediators in the most literal sense of the term insofar as they

---

[27] Agricola, xxx-xxxi.

[28] Charles Whitney, *Francis Bacon and Modernity* (New Haven: Yale University Press, 1986), 112.

> intercede between the external world and the subjective experience of the human intellect or senses.[29]

Wolfe views the machine as an intermediary between objective experience and perception, and, at the same time, as a physical manifestation of that irreconcilable gap. Because the machine "mimics, rather than corrects, the errors and vanities of our sensual instruments," it exposes the fundamental "uncertainties of human experience."[30] Wolfe identifies an ambivalence in Bacon's use of mechanical imagery that can also be detected in his metaphors of mining.

When Bacon, in *The Advancement of Learning*, presents a preliminary outline of his natural philosophy, he employs the figure of the mine to convey both the specific process of his scientific method and also the subtlety that is required of those who practice it:

> If then it bee true that *Democritus* sayde, *That the truthe of Nature lyeth hydde in certaine deep Mynes and Cauves*; And if it bee true likewise, that the *Alchymists* doe so much inculcated, That *Vulcan* is a second Nature, and imitateth that dexterouslie and compendiouslie, which Nature worketh by ambages, & length of time, It were good to deuide Naturall Phylosophie into the Myne and the Fornace, and to make two professions or occupations of Naturall Philosophers, some to be Pionneers, and some Smythes, some to digge, and some to refine and Hammer. (*AL*: 80)

Bacon perceives the secrets of nature in terms of gemstones hidden in mines and caves. He suggests that the practice of mining may be additionally suited to plumbing the depths of nature. He compares the experimenter to the miner and the theoretician to the smith. Just as he forges his own theories, grounded in the observations of Agricola, he encourages all miners of experience to deposit their bounty in the furnaces of the philosopher/smiths. Bacon asserts that one need not

---

[29] Jessica Wolfe, *Humanism, Machinery, and Renaissance Literature* (New York: Cambridge University Press, 1994), 4.

[30] Ibid, 5.

be skilled at both mining and refining; it is only necessary to recognize the distinction between observation and theory, and to be aware of the danger of theory uninformed by observation. Later, Bacon alludes to this extended metaphor when he voices his desire to see nature "forced out of her natural state and... pressed and molded" (*NO*: 25). Like the process of refining metal—first retrieving ore from the earth and then rendering it functional in a furnace—all nature must be subjected to rigorous experimentation, followed by a methodology informed by "instruments and other aids" *(NO*: 45). Then, Bacon predicts, "will we pass from Vulcan to Minerva," from the experience of the underground to exalted knowledge (*NO*: 141). Because Bacon's scientific method is predicated on a reversion to the origin of experience, he is unable to depend on a preexisting scientific framework. Ultimately, metaphor—and metaphor alone—provides the structural basis for his new theory.

In most cases, philosophical writing distinguishes itself from other genres by its emphasis on rational content, while its figurative content is relatively ignored. In *The Philosophical Imaginary*, Michèle Le Doeuff argues that the figures contained within philosophical writing express important theoretical tensions that cannot otherwise be communicated. He explains:

> The meaning conveyed by images works both for and against the system that deploys them. *For*, because they sustain something which the system cannot itself justify, but which is nevertheless needed for its proper working. *Against*, for the same reason—or almost: their meaning is incompatible with the system's possibilities.[31]

Le Doeuff asserts that these images must be interpreted individually before they can be understood within the context of the philosophical system that contains them. Armed with interpretation, Le Doeuff might contend, readers can "pass

---

[31] Michèle Le Doeuff, *The Philosophical Imaginary* (Stanford: Stanford University Press, 1989), 3.

from Vulcan to Minerva," transforming themselves from those who process knowledge to those who produce enlightenment.

In Bacon's utopian fable, *New Atlantis*, voyagers discover an island that epitomizes the production of enlightenment.[32] Unfettered by the constraints of scientific theory, Bacon conjures up, in the words of Donald Worster, "a manmade paradise... rendered astonishingly fertile by science and human management."[33] It is revealing that in Bacon's description of the "manmade paradise" at the island's intellectual center, called Salomon's House, he glosses over what could be considered "astonishingly fertile" gardens and forests. Instead, he begins with a description of underground caves:

> Those caves we call the lower region, and we use them for all coagulations, indurations, refrigerations, and conservations of bodies. We use them likewise for the imitation of natural mines, and the producing also of new artificial metals by compositions and materials which we use, and lay there for many years. We use them also sometimes (which may seem strange) for the curing of some diseases, and for the prolongation of life, in some hermits that choose to live there, well accommodated of all things necessary, and indeed live very long; by whom also we learn many things. (*NA*: 321)

The caves are symbols of Bacon's goal of mastery over mining, and accordingly, of mastery over nature. By creating manmade mines and artificial metals, the men of Salomon's House demonstrate that they have learned to imitate nature, and have succeeded in the symbolic challenge of moving from Vulcan to Minerva that is posited in *The Advancement of Learning*.

---

[32] The fable, composed in 1624 but left unfinished, was published after Bacon's death in 1627.

[33] Donald Worster, *Nature's Economy: The Roots of Ecology* (San Francisco: Sierra Club, 1977), 30.

The goal of a transition from Vulcan to Minerva, god to goddess, is significant, since women are notably absent from Salomon's house. The "fellows" of Salomon's House are celibate (*NA*: 329). The only women who appear inside the vaunted institution are "servants and attendants" (*NA*: 330). This male bias has been hotly debated. In *Reflections on Gender and Science*, Evelyn Fox Keller explains how Bacon's desire to break from other, ineffectual models of scientific discovery led to a "reliance on the language of gender" that suffused his writing with metaphors of sexual potency, domination, and control.[34] According to Keller, Bacon "sought a philosophy that deserved to be called 'masculine,' that could be distinguished from its ineffectual predecessors by its 'virile' power, its capacity to bind Nature to man's service and make her his slave."[35] Bacon's description of the caves of Salomon's house would seem to support this assertion. The caves "imitate" natural mines, replacing the problematic conception of the mine-as-womb with a genderless image. They generate metals not by natural methods, but by "artificial" means. In the caves of Salomon's house, the role of women has been replaced by technology—and this is a tremendous benefit: the hermits who choose to live there experience a "prolongation of life." Without the threat of mortality, women become further reduced in value, since their reproductive capacity is no longer essential to the survival of the species.

Keller describes the evolution of Bacon's conception of natural philosophy in terms of a "complex sexual dialectic."[36] She argues that "beyond the overt insistence on the virility and masculinity of the scientific mind lies a covert assumption and acknowledgement of the dialectical, even hermaphroditic, nature of the 'marriage between Mind and Nature.'"[37] What does this conception

---

[34] Evelyn Fox Keller, *Reflections on Gender and Science* (New Haven: Yale University Press, 1985), 7.

[35] Ibid., 7.

[36] Ibid., 35.

[37] Ibid., 40.

explain about the metaphor of mining? To begin with, the caves of Salomon's House represent a departure from reality. The realm of the underworld, usually associated with the end of life, is linked here to its "prolongation." While the darkness of caves generally impedes discovery, in Bacon's fable, the resident hermits find enlightenment. Why does Bacon impose this artifice on his conception of the cave? The unlimited potential of the manmade cave reinforces the broad goals of Bacon's scientific exploration: his desire to "enlarge the bounds of human empire, to the effecting of all things possible" (*NA*: 321). Perhaps by locating his most ambitious endeavors below the earth, Bacon believes he can limit possible accusations of audacity and hubris.[38]

In his philosophical treatise, Bacon's description of the cave is less "astonishingly fertile," and more indicative of the practical considerations of a manmade environment than his critics acknowledge. In the preface to *The Great Instauration*, Bacon strains to convey the pathos of his solitary scientific mission. He explains, "I have committed myself to the uncertainty, difficulties and loneliness of the ways and, trusting in God's help, have held my mind steady against the violent forces and serried ranks of opinion" (*NO*: 13). As Mumford characterizes the miner as involved in "dogged, unremitting, concentrated work," the image that Bacon evokes is that of a solitary journeyman, resolute in the face of adversity, yet resigned to the "unremitting" nature of his own mental exertions.[39] Julie Solomon, in *Objectivity in the Making*, identifies a commonality among the "self-distancing, prudential strategies of [the] vocational travelers" that Bacon employs to illustrate the scientific practices of *The Great Instauration*.[40]

---

[38] Interestingly, Bacon considers most men already too brazen. In *New Organon*, he writes, "It is not feathers we must provide for the human understanding so much as lead and heavy weights, to restrain all the leaping and flying" (111).

[39] Mumford, 70.

[40] Julie Solomon, *Objectivity in the Making: Francis Bacon and the Politics of Inquiry* (Baltimore: Johns Hopkins University Press, 1997), xviii.

Each of these men—the merchant, the miner, the diplomat, and the spy—rely on objective, mechanical processes in order to complete their work. For Solomon, Bacon's innovation is his emphasis on the process of "self-distancing" that produces scientific objectivity. The mine, as an enforcer of this objectivity, could be interpreted as an emblem of the accuracy and precision of Bacon's new science.

On the other hand, the mine can also stand for the danger of an objectivity that is taken too far. Mumford characterizes the mine as "a dark, a colorless, a tasteless, a perfumeless, as well as a shapeless world: the leaden landscape of a perpetual winter."[41] He conceives of the mine as a barren expanse, a representation of a world in which the richness of subjective experience is replaced by the emptiness of experimental science. Rosalind Williams also conceives of the mine in terms of "displacement."[42] She foresees consequences much like the conditions that Mumford describes. According to Williams, "Nature subjected to scientific rationalism... ceases to be a vital source of human meaning."[43] It is true that Bacon campaigns for scientific rationalism at the expense of sensory experience. His insistence that the "subtlety of experiments is far greater than that of the sense itself" suggests a strong preference for objective thought (*NO*: 22). In his statement, however, he acknowledges that "sense" cannot be abandoned altogether. The human senses provide the tools for assessing the validity of scientific experiments. Furthermore, the knowledge that stems from experimentation ultimately enhances perception and the understanding of the natural world. Bacon aspires not to a world in which sensory experience is absent, but to a laboratory in which sensory experience can be studied and contained. For

---

[41] Mumford, 70.

[42] Williams, 4.

[43] Ibid., 22.

this reason, the mine emerges as an ideal environment for the observation of, and experimentation on, nature.

Nevertheless, it is the absence of sensory experience within the mine that contributes to making it such an interesting structural device. Whereas the uneven surfaces of the earth "entirely lack any unblemished and smooth surface on which to receive the true rays of things," the bare walls of the mine act as a blank canvas, allowing those who venture underground to project onto the earth their literal (and alliterative) mines of the mind (*NO*: 23). Mumford positions the mine as the "first completely inorganic environment to be created and lived in by man."[44] He explains:

> Except for the crystalline formations, the face of the mine is shapeless: no friendly trees and beasts and clouds greet the eye. In hacking and digging the contents of the earth, the miner has no eye for the forms of things: what he sees is sheer matter, and until he gets to his vein it is only an obstacle which he breaks through stubbornly and sends up to the surface. If the miner sees shapes on the walls of his cavern, as the candle flickers, they are only the monstrous distortions of his pick or arm: shapes of fear. Day has been abolished and the rhythm of nature broken: continuous day-and-night production first came into existence here. The miner must work by artificial light even though the sun be shining outside; still further down in the seams he must work by artificial ventilation, too: a triumph of the "manufactured environment."[45]

Mumford's depiction of the "manufactured environment" is disheartening; but a noteworthy aspect of his description is its emphasis on the "monstrous distortions" and the "shapes of fear," the underworld of illusions projected into the mine by the miners themselves. These images resonate in Bacon's writing. He, too, speaks of the "uncertain light of the sense" that reveals "ambiguous paths" (*NO*: 12). Although, presumably, Bacon never entered a mine, he is like a miner. He is impelled to "hold steady" in face of "inward hesitations and scruples,

[44] Mumford, 68.

[45] Ibid., 69-70.

no less than against the dark clouds of Nature and the phantoms flitting around me on every side" (*NO*: 13).[46]

These shadows and phantoms recall Plato's allegory of the Cave, in which prisoners held underground mistake the shadows of puppets projected onto the cave walls for the figures of real people in the world above. Arendt argues that Plato's Cave is a revision of prior conceptions of the subterranean world. She writes, "The senseless, ghostlike motion ascribed by Homer to the lifeless existence of the soul after death in Hades is now ascribed to the senseless doings of men who do not leave the cave of human existence to behold the eternal ideas visible in the sky."[47] Bacon would most certainly agree with this assessment of the "senseless doings of man," but yet another reversal is required in order to apply Arendt's interpretation to Bacon's scientific theory: man must give up the search for knowledge in the heavens and instead turn back to the bowels of the earth.[48]

Solomon offers a different assessment of Bacon's use of Plato's allegory of the Cave. Historically, she explains:

> To dwell in the cave of the Platonic "idols" was to reside in the distorting cave of the physical and the sensual... and to be deprived of direct access to the immaterial world of truth. Only when we get to Bacon's era does the focus narrow, so that the "idols of the cave" become the idols of idiosyncrasy that Bacon

---

[46] As in the dark passages of the mine, Bacon gives voice to the fear that men can also lose their way in the "labyrinth" of experience (*NO:* 13). See also Bacon's *Wisdom of the Ancients* for an interesting gloss on the Greek myth (Kila: Kessinger, 1997).

[47] Arendt, 265-6.

[48] Arendt argues that the major philosophical movement of Bacon's age was the reordering of action over contemplation. She summarizes, "The fundamental experience behind the reversal of contemplation and action was precisely that man's thirst for knowledge could be assuaged only after he had put his trust into the ingenuity of his hands... In order to be certain one had to *make sure*, and in order to know one had to do." Ibid., 207.

> deplores: "every one philosophizes out of the cells of his own imagination, as out of Plato's cave."[49]

Solomon describes the shift in the interpretation of Plato's allegory in terms of a narrowing of "focus." It would seem that the mine, which focuses beams of light through narrow mineshafts, could provide an updated model for the exploration and analysis of individual idiosyncrasies. Separated from cultural and religious influences, the mine remains impenetrable to the light of heavenly wisdom. In the darkness, each man can process the knowledge of nature according to his own method. And unlike the cave, which imprisons man in both literal and figurative ways, the mine provides a means of egress. Once man has identified and isolated the "idols" of his mind, he can rise to the surface to rejoin—and enlighten—his fellow men.

In *Technics and Civilization*, Mumford asks, "Did the mine acclimatize us to the views of science? Did science in turn prepare us to accept the products and environment of the mine?"[50] Indeed, the method of scientific discovery that Bacon promulgates in *The Great Instauration* ushers in a new era of science. His emphasis on exploration, of the importance of theory informed by observation, revolutionizes the perception of science in pre-Restoration England. Bacon's use of the metaphor of the mine contributes additional depth to his critique and revision of the scientific method. By relocating the source of knowledge from the heavens to the underground, Bacon challenges many of the cultural prohibitions that had limited subterranean exploration. Bacon's revised scientific topography reveals new recesses for intellectual inquiry as well, and anticipates the Freudian topographical model of the mind. Bacon's reliance on the metaphor of mining, as demonstrated by his commitment to the unearthing of truth, provides a compelling historical example of the interrelation of metaphor and scientific theory.

---

[49] Solomon, 23.

[50] Mumford, 70.

Lauren Klein (City University of New York) is a doctoral candidate in the Department of English at the City University of New York. Her research interests coalesce around representations of technology in 19th century American literature. She has taught literature and composition at Brooklyn College and currently works as an Instructional Technology Fellow at Queens College. She is the author, most recently, of "Walking and Writing the High Line" (2007).

Gender and Scientific Authority: Huarte de San Juan's *Examen de ingenios* and Oliva Sabuco's *Nueva filosofía*[1]

Beatriz Cruz-Sotomayor

> *. . . no woman came to the world with a wit apt for knowledge . . . . Therefore we are to shun this sex, and to procure that the child be borne male: for in such only resteth a wit capable of learning.*
>
> Huarte de San Juan[2]

> *. . . [and I pray that] your Majesty will receive this book by a woman, the quality of which I think is better than that of those by men [whether they be] vassals or great lords. This book is as unusual and novel as its Author.*
>
> Oliva Sabuco[3]

When the *Examen de ingenios* (*Examination of Men's Wit*) by Doctor Juan Huarte de San Juan is looked at as the immediate scientific predecessor of Oliva

---

[1] This paper was translated in part by Emily J. Sutherland.

[2] ". . . ninguna mujer nace con ingenio y saber . . . . Por tanto se debe huir de este sexo, y procurar que el hijo nazca varón, pues en él solo se halla el ingenio que requieren las letras," Juan Huarte de San Juan, *Examen de ingenios* (Madrid: Biblioteca de Autores Españoles. Obras Escogidas de Filósofos, 1953), 497. Quotes in English are taken from the facsimile edition: *Examen de ingenios. The Examination of Men's Wits (1594) by John Huarte,* trans. M. Camillo Camilli and R.C. Esquire (Gainesville: Scholars' Facsimiles & Reprints, 1959), 287. On subsequent footnotes I include the Spanish original texts of both Huarte and Sabuco's quotations.

[3] ". . . y reciba V. Magestad este servicio de una muger, que pienso es el mayor en calidad, que quantos han hecho los hombres, vassallos, ò señores . . . . Tan extraño y nuevo es el libro, quanto es el Autor," Oliva Sabuco de Nantes, *Obras de doña Oliva Sabuco de Nantes y Barrera (Escritora del siglo XVI)*, ed. Octavio Cuartero (Madrid: Establecimiento Tipográfico de Ricardo Fe, 1888), XLI-XLII. On subsequent references I will refer to this edition as *Nueva filosofía.* The English translations are mine. I acknowledge the generosity of Mary Ellen Waithe on the revision of my translation of the *Nueva filosofía*'s quotations. The first English translation of the *Nueva filosofía* is now in press: *New Philosophy of Human Nature Neither Known to Nor Attained by the Great Ancient Philosophers, Which Will Improve Human Life and Health,* trans. and eds. Mary Ellen Waithe, María Colomer Vintró, and Ángel Zorita (Illinois: University of Illinois Press, 2007).

Sabuco's *Nueva filosofía de la naturaleza del hombre* (*New philosophy of Human Nature*), it unveils a problem that deals not only with the scientific principles on the psycho-physiology of women, but also with what has been said about female authorship and scientific authority.

Dr. Huarte's influential treatise about men's wit was published in 1575 in Madrid. It deals with theories on humoral disposition in order to explain and define not only the superiority of some men over others, but also the superiority of men over women. He emphasizes the limited intellectual capacity of women which, according to humoral theory, is due to their cold and moist quality. The scientific misogyny of Huarte's work echoes the claims of Aristotle and Galen. His emphasis on the natural indisposition of women for wit is so strong that Huarte even explains a procedure of six steps or *diligencias* in order to conceive males and not females during sexual intercourse.[4] Let's look at the beginning of his discussion on this topic:

> Those parents who seeke the comfort of having wise children, and such as are towards for learning, must endevour that they may be borne male: for the female, through the cold and moist of their sex, canot be endowed with any profound iudgment. Only we see, that they talke with some apparence of knowledge, in slight and easie matters, with termes ordinary, and long studied, but being set to learning, they reach no farther than to some smacke of the Latine tongue, and this only through the help of memorie.[5]

---

[4] Huarte exposes his six steps as follow: "One of which is, to eat meats hot and drie. The second, to procure that they make good digestion in the stomacke. The third, to use much exercise. The fourth, not to apply themselves unto the act of generation, untill their seed be well ripened and seasoned. The fifth, to companie with the wife foure of five daies before her naturall course is to runn. The sixth, to procure, that the seed fall in the right side of the womb, which being observed (as we shall prescribe) it will grow impossible, that a female should be engendred." (*Examination of Men's Wit*, 289) ("Una de las cuales es comer alimentos calientes y secos. La segunda procurar que se cuezan bien en el estomago. La tercera hacer mucho ejercicio. La cuarta no llegarse al acto de la generacion hasta que la simiente esté cocida y bien sazonada. La quinta tener cuenta con su mujer cuatro ó cinco dias ántes que le venga la regla. La sexta procurar que la simiente caiga en el lado derecho del útero. Las cuales guardadas, como dirémos, es imposible engendrarse mujer." (*Examen de ingenios*, 497).

[5] *Examination of Men's Wit*, 286. "Los padres que quisieren gozar de hijos sabios y que tengan habilidad para letras, han de procurar que nazcan varones, porque las hembras, por razón de la frialdad o humedad de su sexo, no pueden alcanzar ingenio profundo; sólo vemos que hablan con alguna apariencia de habilidad en materias livianas y fáciles, con términos comunes y muy

Oliva Sabuco's text defies Huarte's medical treatise in many ways. The *Nueva filosofía,* published in 1587—twelve years after *Examen de ingenios*—offered a new natural philosophy, told in a female voice, challenging, at first, the concept of the limited intellectual capacity of women, perpetuated, as we have just said, by Huarte de San Juan.

The text signed by Doña Oliva Sabuco appeared in Madrid under the explicit title *Nueva filosofía de la naturaleza del hombre, no conocida ni alcançada de los grandes filosofos antiguos: la qual mejora la vida y salud humana* (*New Philosophy of the Human Nature not Known and not Reached by the Ancient Philosophers, that Improves Human Life and Health*)[6]. The book is structured in the form of dialogues among three shepherds, and occasionally, between a doctor and a shepherd. It consists of seven books: five in Spanish and two in Latin.[7] These books are preceded by the table of contents, a letter where Oliva requests the favor and protection against the detractors of her book, a prologue to the reader, a dedication letter to the king Phillip II, two sonnets by Juan de Sotomayor praising Oliva and her work, and official documents for publication.

These dialogues are grounded on the notion of the *Nosce te ipsum.* With the understanding of their own nature, humans are able to preserve their health and prolong their lives. Thus, the text provides a natural philosophy—essentially psychosomatic—based on the direct relationship between emotions and physical health. To propose this relationship between the mind and the body, Sabuco suggests that the brain controls all bodily functions. Although she makes

---

estudiados, pero metidas en letras no pueden aprender más que un poco latín, y esto por ser obra de la memoria." (*Examen de ingenios,* 497).

[6] This is the English title according to the cited English edition.

[7] These are: *Colloquy on Human Nature and Knowledge of One's Self*; *Treatise on the Composition of the World*; *Treatise about Things which will improve the World*; *Colloquy on Remedies*; *Colloquy on Errors of traditional Medicine*; *About Human Nature*; and, *True Philosophy*. These English titles are taken from Waithe and Vintró's Web site on Oliva Sabuco: http://www.sabuco.org.

reference to rudimentary knowledge of humoral theory, she believes human life depends on a substance that resides in the brain. A white sap, called *chilo*[8], whose rise (*cremento*) or fall (*decremento*), caused by internal and external motivations, determines the health or sickness of a human being.

It is important to point out that the *Nueva filosofía* is faced with a fiery controversy with regard to its author. Although it had already been questioned whether Oliva Sabuco was the true author, it was not until 1903 that a property registrar from Alcaraz found the will of Miguel Sabuco, Oliva's father. In it, he says that he is the true author of the medical text, not his daughter.[9] Nevertheless, what interests us at this point is not the true author of the text, but rather the presence of a woman's name on a scientific text, the strategies used to give authority to this female voice, and the way the woman is handled scientifically—as opposed to the way Huarte handles this subject.

Sabuco's dedication letter to King Phillip II opens with the undeniable presence of a female voice: "[I am] a humble woman-servant and vassal, kneeling

---

[8] The word *chilo* or 'chyle' derives from the late Latin *chylus*, from the Greek word *khylos* which means 'juice.' Nowadays it is the lymph that is milky from emulsified fats. It is believed to be used by Hippocrates and later elaborated by Galen to designate the juice produced by digestion and converted into blood in the liver. For Sabuco it is a more complex substance which originates from digestion but at the brain from where it spreads through the nerves throughout the whole body. The primary role of the brain, its *chilo,* and the nerves as transmitters set a new conception of the brain and its functions near modern neurophysiology. See: *Webster's New Twentieth Century Dictionary. Unabridged*, Second Edition (William Collins Publishers, Inc., 1980), 324; Luis S. Granjel, "El jugo nérveo," in *La doctrina antropológico-médica de Miguel Sabuco* (Salamanca: Universidad de Salamanca, 1956), 43-50; A. Martín-Araguz, C. Bustamante-Martínez, and V. Fernández-Armayor, "El *suco nérveo* sabuceano y los orígenes de la neuroquímica en el Renacimiento español", *Revista de Neurología* 36 (2003): 1190-1198.

[9] For more on the issue of authorship see: Fernando Rodríguez de la Torre, "El autor y la autoría en la obra de Sabuco", *Al-Basit. Revista de Estudios Albacetenses. Monográfico dedicado a Miguel Sabuco* 22: (Dec. 1987): 233-265; Damaris Otero Torres, "'Una humilde sierva osa halbar' o la ley del padre; dislocaciones entre texto femenino y autoría masculina en 'La carta introductoria al Rey Nuestro Señor' de Oliva Sabuco de Nantes", *Taller de Letras* 26: (1998): 9-27; Beatriz Cruz-Sotomayor, "Una voz en busca de autorización: Oliva Sabuco y su filosofía médica de la felicidad" (M.A. thesis, University of Puerto Rico, 2001). For my PhD dissertation I am currently working extensively on the hermeneutical implications of the polemical authorship of the *Nueva filosofía.*

in absence [because I] cannot dare to speak in presence . . . ."[10] The beginning of the "Letter to the King" deals with the issue of a humble woman who dares to speak. A woman writer was supposed to upset natural order, but in Oliva's text, the justification for this daring act is founded on the same ideas that reinforce her natural subordinate position in relation to masculine superiority. She argues:

> That great law of ancient chivalry . . . gave me this courage and boldness. [That law of chivalry] was to always favor women in their endeavors. The natural law of magnanimous generosity which always favors the weak and humble just as it destroys the arrogant also encouraged me. The grand, unlearned magnanimity of the Lion, king and lord of all animals, shows mercy towards children and weak women, especially when [the latter], prostrate on the ground [before the Lion] and has the audacity and [summons the] strength to speak. Likewise, here, I, with such boldness and courage dare to offer and dedicate my book to Your Majesty and request favor of the great Lion, King and Lord of all men in asking for the protection and shelter of the aquiline wings of our Majesty, under which [wings] I place this book, my son, whom I have engendered.[11]

Thus, the right to speak is granted by the "humble" and "weak" nature of the female gender. The feminine scientific discourse of Sabuco, strategically gains its authority by emphasizing gender itself. Sabuco acquires a valid textual voice as a woman who dares to write and who deserves the protection of the powerful. The female authorship of the *Nueva filosofía* threatens the old traditional theories

---

[10] "Una humilde sierva, y vassalla, hincadas las rodillas en ausencia, pues no puede en presencia, osa hablar . . . ." (*Nueva filosofía,* XLI).

[11] "Diòme esta osadia, y atrevimiento aquella ley antigua de alta cavalleria, à la qual los grandes señores, y Cavalleros de alta prosapia, de su libre, y espontanea voluntad, se quisieron atar, y obligar, que fuè favorecer siempre à las mugeres en sus aventuras. Diòme tambien atrevimiento aquella ley natural de la generosa magnanimidad, que siempre favorece à los flacos, y humildes como destruye à los sobervios. La magnanimidad natural, y no aprendida del León (Rey, y señor de los animales) usa de clemencia con los niños, y con las flacas mujeres, especial si postrada por tierra tiene osadia y esfuerzo para hablar, como tuvo aquella cautiva de Getulia, huyendo del cautiverio por una montaña, donde avia muchos leones, los quales todos usaron con ella de clemencia, y favor, por ser mujer, y por aquellas palabras que osò decir con gran humildad. Pues assi yo con este atrevimiento, y osadia oso ofrecer, y dedicar este mi libro à V. Catholica Magestad, y pedir el favor del gran Leon, Rey, y señor de los hombres, y pedir el amparo, y sombra de las Aquilinas alas de V. Catholica Magestad, debaxo de las quales pongo este mi hijo, que yo he engendrado." (Ibid.).

of sexual differentiation based on the natural intellectual inferiority of women. In the same line of thought as ancient philosophers and the so called Fathers of Christianity, Huarte states about women:

> . . . in her naturall disposition, all sorts of learning and wisdome, carrieth a kind of repugnancie to her wit. And for this cause, the Catholicke Church, upon great reason hath forbidden, that no woman do preach, confesse, or instruct: for their sex admitteth neither wisdome nor discipline.[12]

Although, in Huarte's work, women are classified according to three levels of coldness and moistness, he states that women are even incapable of surpassing the wit of Eve, who, furthermore, possessed "less knowledge than Adam." Only if a woman has "any gift from above," beyond her natural disposition she "may well teach and speake."[13] For Huarte, the abundance of wise women in Greece is a rare and dreadful phenomenon. He assures:

> But which we may most marvell[14] at in Greece is, that wheras the wit of women is found so repugnant unto learning (as hereafter we will prove) yet there have been so many she Greekes, so specially seen in the sciences, as they have grown into competencie with the sufficientest men: as namely Leontia, a most wise woman, who wrote against Theophrastus, the greatest Philosopher of his time, reprooving him for many errors in philosophy. But if we looke into other Provinces of the world, hardly shall we find sprung up any one wit that was notable.[15]

---

[12] *Examination of Men's Wit*, 275. ". . . quedando la mujer en su disposición natural, todo género de letras y sabiduría es repugnante à su ingenio. Por donde la Iglesia católica con gran razon tiene prohibido que ninguna mujer pueda predicar, ni confesar, ni enseñar, porque su seso no admite prudencia ni disciplina." (*Examen de ingenios,* 493-494).

[13] Ibid., 274 and 275.

[14]The word 'marvell' has been used to translate the Spanish word 'espanta'. As stated in etymological dictionaries of the Spanish language, the verb *espantar* derives from the Latin '*expavere*' which means 'to cause fear, horror or astonishment.' The negative implications of the verb used by Huarte go beyond the neutral notion of 'marvell.' Cf. Joan Corominas and José A. Pascual, *Diccionario Crítico Etimológico Castellano e Hispánico* (Madrid: Gredos, 1984), 734; María Moliner, *Diccionario del Uso del Español* (Madrid: Gredos, 1998), 1196; Sebastián de Covarrubias, *Tesoro de la Lengua Castellana o Española*, ed. Martín de Riquer (Barcelona: S.A. Horta, 1943), 551.

[15] *Examination of Men's Wit*, 241. "Pero lo que más espanta de Grecia es, que siendo el ingenio de las mujeres tan repugnante á las letras, como adelante probarémos, hubo tantas griegas

Huarte's insistence on female's inability to deal with scientific and philosophical issues—perpetuated in Renaissance Spain not only by him but by many other influential thinkers such as Juan Luis Vives, Luis de León and Francisco de Mello—clashes with Sabuco's allegation of the importance of her natural philosophy. Sabuco assures that her book must be considered the best of them all. She thus discredits the great medical and philosophical personalities of the past. She alleges:

> This book was just as wanting in the world as others were unnecessary. Galen, Plato and Hippocrates needed this book in their treatises on human nature. Aristotle needed it too when he discussed the soul, life and death. The natural philosophers like Pliny, Aelian and others also needed this book when discussing human nature. This was the most needed, best and most fruitful philosophy for mankind, but it was left untouched upon by the great ancient philosophers. . . . but it is as clear and evident as sunlight that ancient medicine, whose main principles are read and studied, is wrong because ancient philosophers and physicians neither understood nor attained [an account of] their own nature, that is, where medicine originates and upon which it is founded.[16]

Ancient medicine and philosophical authorities are clearly displaced by that of Sabuco. Might this also be a rejection of those scientific authorities that perpetuated the principles of a female physiology that was cold and moist, and therefore inferior, as opposed to a physiology based on the rise and fall of a

---

y tan señaladas en ciencias, que vinieron á competir con los hombres muy racionales, como se lee de Leoncia, mujer sapientísima, que siendo Theophrasto el mayor filósofo que hubo en su tiempo, escribió contra él, notándole muchos errores en filosofía. Y si miramos las otras regiones del mundo, apénas ha salido de ellas un ingenio que sea notable." (*Examen de ingenios,* 484).

[16] "Este libro faltaba en el mundo, assi como otros muchos sobran. Todo este libro faltò à Galeno, à Platon, y à Hipocrates en sus tratados de natura humana, y à Aristoteles quando tratò de *anima y de vita & mortae.* Faltò tambien à los naturales, como Plinio, Eliano, y los demàs, quando trataron de *homine.* Esta era la philosofia necessaria, y la mejor, y de mas fruto para el hombre, y esta toda se dexaron intacta los grandes Filosofos antiguos. . . . pero resulta muy clara, y evidentemente, como resulta la luz del Sol, estàr errada la medicina antigua, que se lee y estudia en sus fundamentos principales, por no aver entendido ni alcanzado los Filosofos antiguos, y Medicos su naturaleza propia, donde se funda, y tiene su origen la medicina." (*Nueva filosofía,* XLII-XLIII).

substance, provoked by immediate external circumstances? This text attempts to be original and different at all costs: "This book is as unusual and novel as its Author," affirms Sabuco.[17]

The female voice that introduces Sabuco's text with the "Letter to the King" immediately dissipates in the pastoral dialogue among Antonio, Veronio and Rodonio. The shepherd Antonio assumes the main voice, while he explains his theories about human health. The female voice of the "humble woman-servant" disappears but it encloses us in a discourse that is no longer in the voice of a woman, but rather on the lips of a simple shepherd, who does not have the authority to address these scientific topics.

Antonio discusses afflictions and their effect on men and women who, with a few exceptions, he groups under the name "men." Afflictions such as anger, sorrow, fear, jealousy, the sound of a harquebus, pride, and arrogance, affect women—especially pregnant women—more than men. Such concepts—which imply sexual differentiation based on the biological essence of women—accentuate the essentialist perspective that is implicit in the dedication letter to King Phillip II: that the biological constitution of women is naturally more vulnerable to certain afflictions and circumstances.

Let us remember that Sabuco's medico-philosophical theory is founded on the brain's state of rise or fall of the *chilo* and not merely on the traditional theories of the four fluids and elements. In Sabuco's text, the hierarchical oppositions between hot and cold, and moist and dry, and their correspondence to female and male, respectively—which has an important role in Hippocratic medicine and in the natural philosophy of Aristotle—is enriched by the opposition between the rise and fall of the white sap, which originates in the brain according to external circumstances and not to natural dispositions.

Also, the white sap is described by Sabuco as the "first" and "principal" matter, of an "active" and "formative" quality, whereas blood is a "secondary matter." Most of the white sap goes down the spinal medulla to the kidneys,

[17] Ibid., XLII.

where it becomes sperm and then enters the seminal veins. Contrary to Aristotle's theories about the refining of blood into a white substance, Sabuco relies on Galen's notion of the *chyle* as the raw material from which blood is made. Sabuco states:

> It is laughable what people say, that sperm and milk are red blood whitened inside their vessels. Don't be afraid of this white brain sap, wait until you see how it functions when it descends into the female's uterus; together with the nurture and sustenance of menstrual blood it creates another whole new animal.[18]

Sabuco laughs about Aristotle's theory, but at the same time, emphasizes the superiority of masculine qualities by attributing to them—just as Aristotle did—the formative and active capacity. She also accentuates this superiority by defining this white substance as a "principal" matter way more important than blood itself. This implies that the qualities of men's sperm should be valued above blood, and, above female menstrual blood, defined by Sabuco as "defluxion of the brain"[19] that "serves as a secondary matter for the [creation of the] species in the uterus."[20] Sabuco also discusses contemporary polemical issues regarding conception. It is in the chapter entitled "Improvements in Marriages and Procreation" that Sabuco deals with the two seeds theory, that is, the idea about the contribution of both male's and female's seeds during conception. Sabuco argues:

> . . . the strength and perfection of humans is neither hereditary nor it is propagated by generation by semen alone, as in plants; in humans the seed of both is needed. If two seeds, male and female, do not conjoin, [semen alone] cannot engender. From this mixture of two things a third being is created which is neither the one nor

---

[18] "Es cosa de risa lo que dicen, que la esperma, y la leche son sangre colorada, y que en sus vasos se buelve blanca: y de esto que he dicho (señor Doctor) de esta sangre blanca, y sus grandes obras, de este jugo blanco del celebro, no os espanteis, pues veis lo que hace caído en el utero de la hembra, que hace de nuevo todo el animal con el riego, y sustento de la sangre del menstruo, que mas es hacer el cuerpo todo de nuevo, que aumentar lo hecho." (Ibid., 281-282).

[19] Ibid., 225.

[20] Ibid., 337.

> the other. This is why we see fools born from the wise, cowards from the strong, timid and pusillanimous sons from magnanimous and heroic men . . . that is why a man should pay attention to relevant [qualities] of the woman who is his prospective mate: in the interests of the quality of their children. Likewise the woman should pay attention to the qualities of the man who is her prospective mate. From two high-quality materials a third, high-quality one results. That is, each mate provides half [the qualities].[21]

We see that she is concerned—as in the Hippocratic and Aristotelian traditions—with establishing a balance between the contribution of the seed of the man and of the woman. The same is true for Huarte, but when it comes to the issue of male or female offspring, the case is different. As we have seen, Huarte strongly stresses the importance of conceiving males and not females. Unlike Huarte, Sabuco does not insist on the generation of male children, but rather, just exposes a variety of believes on this matter:

> Some philosophers said that the cause of the generative act is the imagination, as well as the stars and signs that govern us during procreation. Others say that with the right testicle males are engendered, and females with the left. What I find to be the truth is that the Sun helps in the generation of males, and the Moon in that of females; and this way, the absence of the Moon and the presence of the Sun, which will be during the summer, will help the masculine gender; and the absence of the Sun and the presence of the Moon, which will be during the winter, and the full moon,

[21] ". . . la virtud, y la perfeccion del hombre no desciende, ni se propaga en su generacion, como en las plantas, porque aqui solamente basta la simiente de uno, y alli es necessaria la simiente de dos, que sino concurren las dos simientes de varon, y muger, no se engendra; y de esta mistura de dos, se hace una cosa tercera, que ni es esta, ni aquella, como de vino, y agua se hace una tercera cosa, que ni es vino, ni es agua; y assi comunmente salen los hijos mezclados, que ni parecen al padre, ni à la madre, aunque algunas veces parecen totalmente al uno, y fuè, porque la simiente de aquel venciò, y prevaleciò mas, y no huvo total mistura, y assi tomò su forma, conforme à la materia que mas prevaleciò, y mas virtud, y cantidad tenia; pero lo mas comun, es salir mezclados. Y por esto vemos de sabios salir tontos, y de fuertes, cobardes, y de magnanimos, y valerosos hombres, salir hijos apocados, y pusilanimos . . . y assi verà el hombre quanto và en la compañera que toma por muger, para la perfeccion de sus hijos. Y la muger quanto và en el compañero que toma por el semejante, que de dos materias buenas resulta tercera buena: pues el compañero, ò compañera, ha de poner la mitad; por lo qual à los hijos comparò Aristoteles, y muy bien à eslabon, que ata la cadena de en medio, porque el padre puso la mitad, y la madre la otra mitad en el hijo, y assi quedan atados el padre, y la madre con los hijos." (Ibid., 197-198).

will help the feminine gender.[22]

Sabuco knows and handles theories that were still the object of strong debates during her time. In short, she identifies herself with the theory regarding the effect of the sun and moon on conceiving males and females, respectively. In her book, we constantly read about the correspondence between moon, mother, milk, nutrition, decrease, moistness, and coldness—a correspondence diametrically opposed to the superior qualities of the sun, associated with masculinity and defined as increase of the *chilo*. Let us see an example:

> . . . the day is hot and dry, and the night moist and cold. The latter makes the decrement and the first the increment. And for this reason in the mornings, with the presence of the Sun, the sick feel better, and at night they get worse because of the decrement, this is why most of them die at night, and the pain increases, and the pulses vary.[23]

Sabuco, thus, agrees with Aristotle's theory of polarities, but with one major difference: her approach is careful when she works directly with gender issues. There is just one exception. In one of the chapters written in Latin, we find the following commentary about menstruation:

> Women's menstruum or its residue, which in some women is always flowing, can become toxic to the man who comes into contact with it through means of his mouth when it ascends and affects his brain in a few hours. This is just like the poison of an animal bite, and rabies. That is why some men suffer syncopes the same day [they have sex with an infected woman] and others

[22] "Algunos Filosophos dixeron, que era la causa la imaginacion en aquel acto, y las estrellas, y signos, que en la genitura predominan. Otros dicen, que con el compañon derecho se engendran los machos, y con el izquierdo las hembras. Lo mas cierto, que yo hallo es, que el Sol ayuda à la generacion de los varones: y la Luna à la de las hembras; y assi la falta de Luna, y presencia del Sol, que serà en verano en conjuncion, ayudarà al genero masculino, y la falta de Sol, y presencia de la Luna, que serà en invierno, y en plenilunio, ayudarà al genero femenino." (Ibid., 199).

[23] ". . . de manera, que el dia es caliente, y seco, la noche humida, y fria. Esta hace el decremento, y aquella el cremento, y por esso à las mañanas, con la presencia del Sol, se mejoran los enfermos, y à la noche se agravan en el decremento, y por esto los mas mueren de noche, y los dolores crecen, y los pulsos se varian." (Ibid., 278).

become weak and unhealthy.[24]

It is surprising to find this type of theory on the poisonous quality of menstrual blood in Sabuco's text, not only because it is a woman who signs the scientific treatise, but because, as we have already seen, the rest of the text lacks openly negative judgments about women or her natural qualities. Statements such as this one remind us of Pliny's arguments regarding the poisonous and noxious effects of menstrual discharge. In his *Natural History* Pliny states:

> It would be a difficult matter to find anything which is productive of more marvellous effects than the menstrual discharge. On the approach of a woman in this state, must will become sour, seeds which are touched by her become sterile, grafts wither away, garden plants are parched up, and the fruit will fall from the tree beneath which she sits. Her very look, even, will dim the brightness of mirrors, blunt the edge of steel, and take away the polish from ivory. A swarm of bees, if looked upon by her, will die immediately; brass and iron will instantly become rusty, and emit an offensive odour; while dogs which may have tasted of the matter so discharged are seized with madness, and their bite is venous and incurable.[25]

It is interesting that these commentaries are completely absent in the Spanish sections of the *Nueva filosofía*. How would the theory of poisonous menstruation be read in plain Spanish, written by a woman? Is it, perhaps, a way to leave "buried" in Latin, a controversial theory of sexual implications?

Contrary to what we read in a text as fundamental and influential as was Huarte's book, Sabuco's medical treatise surprises us by the subtlety with which it handles the most widespread notions about physiology and gynecology. As we have seen, the textual strategies employed by the author of the *Nueva filosofía* are

---

[24] "Menstruum mulieris, vel eius reliquiae, in aliquibus semper destilantibus, tactu osculi, & coitus, venenum est homini, post aliquot horas ascendendo, & tangendo eius cerebrum, ut venena morfus animalium, & rabies: ideò aliqui sincopas eodem die paciuntur, alii valetudinari afficiuntur." (Ibid., 364). A Spanish translation of Sabuco's Latin books can be found at Samuel García Rubio, "La obra latina del bachiller Sabuco," *Al-Basit. Revista de Estudios Albacetenses. Monográfico dedicado a Miguel Sabuco* 22 (Dec. 1987): 224.

[25] *The Natural History of Pliny*, trans. John Bostock and H.T. Riley, II (London: Henry G. Bohn, 1855), 152.

ambiguous. The text moves between subversion and submission. Female authorship, which is introduced with surprising authority before the King in the dedication letter, cautiously hides behind the voice—also marginal—of the shepherd Antonio. That same way, the original theory of the brain's white sap mingles with the most misogynistic traditions of scientific writing, from Aristotle to her contemporary Huarte de San Juan. The female voice and the pastoral colloquies of the *Nueva filosofía* set a new kind of intertextual dialogue with powerful scientific authorities. Authorship, gender and authority clash, in a unique way, in these two texts of Renaissance Spain.

Beatriz Cruz-Sotomayor (University of Puerto Rico) is a PhD candidate in the Department of Hispanic Studies at the University of Puerto Rico, Río Piedras. She is specializing in Spanish Golden Age literature, Moorish literature, and Renaissance scientific prose and hermeneutics. Currently she is working on her dissertation on the scientific text *New philosophy of the Human Nature* (1587) and the hermeneutical implications of its debated authorship.

Building Gender Into the Elizabethan Prodigy House

Sara L. French

Social and political structures in Early Modern England controlled the understanding of both class and gender in addition to the construction and use of domestic and public space. The spatial divisions of the Early Modern country house—its political, social and cultural imperatives—were maintained and upheld by gender divisions. The category of "woman" supported an ideology that separated men and women as well as separate classes within the domestic environment. This separation and the spatial ideology of class and gender also provided one woman the tools to break down the divisions between men and women, to reinforce class status, and to alter one of the most profoundly masculine spaces within the Early Modern domestic environment—the great hall.

Bess of Hardwick, born about 1527 to a Derbyshire family of the minor gentry, was the patron of at least two major country houses: Chatsworth and Hardwick Hall. She attended the courts of Henry VIII, Edward VI, Mary I, and Elizabeth I and was familiar with the buildings of these monarchs. She also was responsible for a number of smaller building projects and was probably influential in the building undertaken by her fourth husband, George Talbot, Earl of Shrewsbury, Earl Marshal of England and jailor of Mary, Queen of Scots.[1]

Bess's construction of Hardwick Hall exemplifies the traditions and continuities of Tudor building practice as well as significant change in the layout and use of the spaces within it. This building encodes social hierarchy in the Early Modern country house in a new way and helps to dismantle traditional notions of space, gender, and power.

---

[1] David N. Durant, *Bess of Hardwick: Portrait of an Elizabethan Dynasty* (London: Weidenfeld & Nicolson, 1977; reprint 1998), *passim*.

In the last half of the sixteenth century, women were placed in significant positions of royal authority and influence. In England, Mary I and Elizabeth I ruled for over fifty years. In Scotland, Mary Stuart's disastrous reign caused intense upheaval and in France, Catherine de Medici exercised considerable influence over her sons, the kings of France. The construction of prodigy houses in England stemmed directly from changes in royal patronage during Elizabeth I's reign; this change in patronage also altered perceptions about the relationships between men and women in the last half of the sixteenth century.

The binary division of the human race into two distinct, mutually exclusive categories of "male" and "female" was so much a part of the Elizabethan world view that political, social, and even architectural ideology was infused with maintaining that difference. In the Early Modern World, the sexual division of space followed a rigorous division: women had their "place," as did men. However, a contemporary reader of Elizabethan architectural history might be surprised to find that domestic architecture was as much a site for debates over gender ideology as public, governmental, or royal building. The ideology of gender infused the debate over new trends in domestic architecture—that is, in the construction of the country houses of the late sixteenth century.

Domestic architecture in this context should not be understood as small scale or unimportant. Elizabeth I's annual progresses dictated that her courtiers, receiving the favor of a visit from their queen, construct or significantly enlarge their own country houses into palatial buildings suitable for her entertainment. Ian Dunlop writes, "Although [Elizabeth] cannot be described as a patron of the art, she did exercise a formative influence on the architecture of the home counties by making these progresses."[2] As courtiers vied for the favor of the queen, their country houses and plans for the queen's entertainment caused intense competition.

---

[2] Ian Dunlop, *The Palaces and Progresses of Elizabeth I* (London: Cape, 1962), 118-119.

The prodigy house, as defined by architectural historian John Summerson, tempers tradition with innovation in order to impress the queen as well as her courtiers. [3] Each of these houses, which include almost all the buildings with which Robert Smythson was associated, shows some novelty, some particular feature intended to catch the attention of the court as well as the ultimate spectator—the monarch.

Wollaton Hall, built by Sir Francis Willoughby outside Nottingham between 1580 and 1588, and Hardwick Hall, built by Bess of Hardwick between 1597 and 1601, are both defined as prodigy houses. Elizabeth I undertook no new royal building; it was left to those who wished to make an impression to construct new houses that could amaze and please their queen.[4]

Wollaton and Hardwick both demonstrate a novel arrangement of rooms within the house as well as decoration on the exterior. The internal spatial complexity of these buildings shows a clear understanding of the roles men and women were to play. At Wollaton, those roles were rigidly reinforced; at Hardwick, the great hall, serving spaces and state rooms were re-arranged, revealing an awareness of entrenched gender ideology and providing a site for its resistance.

Thomas Laquer's ground-breaking work on the history of gender ideology, *Making Sex: Body and Gender from the Greeks to Freud*, describes the historical process of creating sex in order to reaffirm and justify political and cultural ideas about humanity. In the pre-modern (or, in Laquer's terms, pre-Enlightenment) world there was only one biological sex—man. Those described as women were simply less perfect forms of men. In this construction, one's gender was neither scientifically nor biologically defined. The body was merely labeled as male or female in order to justify its position on the socio-political

[3] John Summerson, *Architecture in Britain 1530-1830*, 9th ed. (New Haven: Yale University Press, 1993), 59-60. Prodigy houses include Theobalds, Burghley, Holdenby, Wardour, Worksop, Hardwick, Longleat, and Wollaton. The last five are all connected to Robert Smythson.

[4] Ibid., 58.

spectrum. If labeled female, it lacked rights, duties, and privileges while at the same time fulfilling the function of bearing children in order to perpetuate the system in which it lived. If labeled male, the body assumed rights, privileges and duties that gave it socio-economic status and ordained that it serve as father rather than mother—as maker rather than bearer of children. Modern notions of scientifically verifiable biological sex are, in this instance, unnecessary, since what is at stake "are not biological questions about the effects of organs or hormones but cultural, political questions regarding the nature of woman."[5]

This one-sex model of gender identity, in Laquer's thesis, permeated the European worldview throughout the ancient and medieval periods. The sixteenth century in England saw its own contributions to the gender debate in the so-called *querelle des femmes,* which centered on women's fitness for rule.[6] But in general, the idea that women were simply less perfect men was the most powerful argument against women's full participation in the political and economic realms.

The presence of Elizabeth I on the throne of England for forty-four years created a unique situation in the Early Modern world. Not only the issue of women as rulers, but the entire ideological basis for kingship was called into question. The king, by nature the father of the country and its people, was in this case a queen: a queen, moreover, whose physical body was small, frail, and feminine.[7] Elizabeth's courtiers were called upon to serve as well as woo and the diplomacy surrounding her possible marriage was extraordinarily complex.[8] But

---

[5]Thomas Laqueur, *Making Sex: Body and Gender from the Greeks to Freud* (Cambridge, MA: Harvard University Press, 1990), 22. This discussion of gender in the Early Modern World is indebted also to the following texts: Sara Mendelson and Patricia Crawford, *Women in Early Modern England, 1550-1720* (Oxford: Clarendon Press, 1998), especially Chapter 2 "Contexts;" Judith Lorber, *Paradoxes of Gender* (New Haven: Yale, 1994); and Mark Breitenberg, *Anxious Masculinity in Early Modern England* (Cambridge: Cambridge University Press, 1996).

[6] Constance Jordan, *Renaissance Feminism: Literary Texts and Political Models* (Ithaca: Cornell University Press, 1990), 86-133.

[7] Ibid., 129-130, 137.

[8] Carole Levin, *The Heart and Stomach of a King: Elizabeth I and the Politics of Sex and Power* (Philadelphia: University of Pennsylvania Press, 1994), 39-64.

more than that, Elizabeth's feminine body was the site for the contest among courtiers for her favor. The favoritism of a king—such as Elizabeth's father, Henry VIII—had been a consistent feature of rule in European countries throughout the medieval period. This contest had been played out among men and in the presence of men alone, in the physical spaces of the king's privy chamber and bedchamber. Access to this privileged space was open only to men—councilors and courtiers who vied for the king's attention.[9]

In every reign, the court had two distinct functions: administrative and ceremonial. The administrative functions were run by the Privy Council and overseen by the monarch. The ceremonial aspects of the monarch's life were organized and run by the royal household. During Elizabeth's reign, these two branches of government were distinct—one run by the male privy councilors, the other run by Elizabeth's ladies-in-waiting and female servants. The women who made up Elizabeth's intimate household held the power to smooth the way into the queen's presence for the men of the Privy Council.[10]

Thus Early Modern gender ideology and real spatial arrangements in Elizabeth's court conflicted: men were in charge of the government, but not in charge of Elizabeth's private life and personal space. This conflict demonstrates that assumptions about space—that men own or organize space and women inhabit it at the pleasure of men—were negotiable in the Early Modern period.[11] Women could and did control their own spaces, excluding or limiting men's access to them. This rift will become even clearer in the analysis of Wollaton and Hardwick.

---

[9] David Starkey, "Intimacy and Innovation: The Rise of the Privy Chamber, 1485-1547," in *The English Court: From the Wars of the Roses to the Civil War,* edited by David Starkey (Harlow, England: Longman, 1987), 71-118.

[10] Pam Wright, "A Change in Direction: The Ramifications of a Female Household, 1558-1603," in *The English Court;* David Loades, *The Tudor Court* (London: B. T. Batsford, 1992), 55-59; 134-136.

[11] Mark Wigley, "Untitled: The Housing of Gender" in *Sexuality and Space*, ed. Beatrize Colomina (Princeton: Princeton University Press, 1992), *passim.*

According to Lena Cowen Orlin, the medieval household was concentrated on the great hall, defined as

> a large room to accommodate the lord's retainers in eating and sleeping, with a central hearth for heating and cooking...Entrance to the house was into a wide corridor known as the screens passage, named for the partial wall or screen that divided this entranceway from the hall.[12]

Several surviving manor houses of the fifteenth century provide examples of the hall and service areas, such as Haddon Hall, Derbyshire; Dartington Hall, Devonshire; and Penshurst Place, Kent.[13]

Haddon's hall is separated from the serving areas by two openings in the screen that provided ceremonial access to the hall for the serving of meals. At Dartington, the screen has three openings. Penshurst retains its original central hearth in the great hall, although the rest of the house has been modified. Most scholars agree that the central hearth began to be replaced by fireplaces in most country houses in the fourteenth century.[14]

Common to all the above examples was the dais placed opposite the screens passage. Meals were served by bringing the food in procession through the screens passage and up to the high table on the dais where the lord sat. This arrangement was continued during the sixteenth century even when the lord was not eating in the hall: meals were ceremonially presented to the lord's place at the high table even if he was absent. The important factor was that the lord maintained his *symbolic* presence as head of the household, as the ultimate authority for those gathered in the hall.[15]

---

[12] Lena Cowen Orlin, "'The Causes and Reasons of All Artificial Things' in the Elizabethan Domestic Environment," *Medieval and Renaissance Drama in England* 7 (1995): 21.

[13] Mark Girouard, *Life in the English Country House: A Social and Architectural History* (New Haven: Yale University Press, 1978), 33-39.

[14] Ibid., 34.

[15] Ibid., 52.

Great houses in England altered their layouts significantly in the fourteenth century and eating habits changed as the nobility moved to the great chamber—usually located on an upper floor beyond the great hall—to dine, listen to music, and entertain important visitors.[16] These changes produced new rooms as well as a staircase at the dais end of the great hall. The ritual of service was extended beyond the great hall, up the stairs, and into the great chamber.

The traditional medieval house has been outlined by architectural historian Mark Girouard in *Life in the English Country House*,

> Changes in the ceremony and hierarchy of eating led to architectural changes. The processional route from the kitchen was often extended up a spacious and richly decorated staircase leading from the hall to the grand chamber, as at Penshurst. The hall tended to get smaller...But it was still the room by which great people entered the house, and its lord and his guests came back into it on great occasions, or on feast days, or for plays. The lavish ceremonial serving of meals in the hall and the generous entertainment of visitors there remained an essential part of the *image* of a great man, even when the upper level of both household and guests had been creamed off into the great chamber. Moreover, it had the weight of *tradition* behind it (emphasis added).[17]

This picture of the ceremony of hospitality emphasizes the important occasion which would certainly have been less formal on ordinary days. It also stresses the way in which the spaces for this ritual were refined over time to codify the hierarchy of space within the great hall. The hall had an "upper" and "lower" end. The upper end was reserved for the dais, great chair, and high table to enhance the lord's position. This description applies to the medieval tradition as practiced by noble landowners into the sixteenth century, even though the monarchy had largely abandoned eating in the great hall by that point. As the

---

[16] C. M. Woolgar, *The Great Household in Late Medieval England* (New Haven: Yale University Press, 1999), 61-63.

[17] Girouard, *Country House*, 52-53.

sixteenth century progressed, the practice of holding elaborate meals in the great hall diminished as a daily occurrence.

Despite the evolution of royal building in the sixteenth century, country houses of the nobility and gentry continued to follow established tradition. Houses such as Haddon, Dartington and Penshurst served as paradigms of medieval ceremonial space. A palace like Hampton Court exercised a profound effect on later builders and fifty year-old models of upper class or royal building were copied by the nobility and gentry. By the end of the century, however, Wollaton and Hardwick each dealt a blow to traditional medieval arrangements, though for distinctly different reasons.

Early Modern England's master builders were not "architects" as Brunelleschi or Bramante were in Italy during the sixteenth century. The lack of architects in the sixteenth century is well-documented, most notably by John Wilton-Ely in his article "The Professional Architect in England." The first person in England to call himself an architect was John Shute, author of a short treatise called *The First And Chief Grounds Of Architecture*, published in 1563, whose patron was the Duke of Northumberland. Shute was well-read in Vitruvius, Palladio and Serlio, but has not been connected with any identifiable building.[18] Only after the death of Elizabeth I and the rise of Inigo Jones in the seventeenth century does the term architect come into regular use by those who had previously occupied the position of "master builder."

Thus, throughout the sixteenth century, the medieval model of craftsmen controlled by a master builder was still the most common system for building in England. Of the half dozen or so named master builders from this period, there exist plans from only a few. One of the most prolific of these was Robert Smythson, who began his career at Longleat, working for Sir John Thynne, in 1568. By 1576, he had accepted a commission from Sir Matthew Arundell at

---

[18] John Wilton-Ely, "The Professional Architect in England," in *The Architect: Chapters in the History of the Profession*, ed. Spiro Kostof (New York: Oxford, 1977, 1986), 180-208.

Wardour Castle. Four years later, Smythson was hired by Arundell's cousin, Sir Francis Willoughby, for his new house at Wollaton.

In the sixteenth century, the phrase "a man's home is his castle" came into general usage well after the decline of the castle as a residence.[19] This phrase would certainly have been popular with a man of the reputation and temperament of Willoughby. His rise in Elizabethan England was due to family connections to the Greys and to his service in the household of Sir Francis Knollys. From there, he rose to prominence and even entertained Elizabeth I at his house at Middleton after her departure from the famous entertainment staged at Kenilworth in 1575 by the Earl of Leicester.[20]

At Wollaton, Sir Francis Willoughby constructed a country house as a showpiece to replace the antiquated manor house in which his family had lived for several decades of the sixteenth century. He sought to create a modern house that still retained the ceremonial and authoritative aspect of the great hall and great chamber. He eliminated the courtyard, instead placing the great hall at the center of the plan, accessed via an oblique route and separated from the passage by an elaborately carved stone screen.

On the exterior, hundreds of windows provide light to the interior and break up the façade; the stone work separating the windows is covered with statues, busts, and elaborate ornamentation. The decorative scheme is drawn largely from pattern books by Dutch architect Vredeman de Vries and the Italian Sebastiano Serlio. De Vries' influence wins, however, in the elaborate strapwork, metopes with grotesque heads and cartouches that cover the façade.[21]

Wollaton's symmetry, on both the inside and the outside, displays the increasing use of inspiration from continental sources. On the interior, the

---

[19] Lena Cowen Orlin, *Private Matters and Public Culture in Post-Reformation England* (Ithaca: Cornell University Press, 1994), 2.

[20] Alice Friedman, *House and Household in Elizabethan England: Wollaton Hall and the Willoughby Family* (Chicago: University of Chicago Press, 1989), 24.

[21] Ibid., 95-96; 108-109.

original entrance (now changed) led to a corridor with two right-angle turns which eventually directed the visitor to the screens passage, then into the hall. The elaborate stone screen with its three large doorways provided access to the hall, with its curious hammerbeam roof and "prospect room" above it. Alice Friedman's reconstruction of the original ground and first-floor plans shows that the arrangement was traditional.[22] The hall rose two stories and had an area devoted to a dais and high table; doors led from the hall to a parlor on the northeast. A staircase led to the south great chamber, with its matching north great chamber on the other side of the hall. The gallery on the northeast side of the house connected the two great chambers. Wollaton's great chamber was located on the first floor above the parlor—as at such traditional houses as Haddon Hall.

The ritual of service at Wollaton was an essential part of Willoughby's search for display and importance that he sought for himself—he was intimately connected to the court but was in fact reluctant to attend the queen. At Wollaton he could be in charge and suffer no one to order his comings and goings, something that was impossible at Elizabeth's court. Friedman writes:

> The household was run with strict attention to protocol and ceremony; a set of regulations written by Francis Willoughby himself describes the service of meals in the hall and great chamber, the rules for receiving visitors, and the tasks of the usher, butler, and under butler. Watching over this small army of officers, servants, visitors, and laborers was Willoughby himself.[23]

These instructions, which might have been written two centuries earlier, show us that Wollaton's building fabric unites with the personality of its owner to reinforce a gender ideology that places the male at the top and the female--Willoughby's contentious and physically frail wife—in a subservient position. Willoughby's household was run by men. His wife was in many ways secondary

---

[22] Ibid., 92.

[23] Ibid., 29.

to the fulfillment of the medieval rituals of lordly display to which Willoughby was attached. His design of the physical spaces at Wollaton reflects his emphasis on customary forms of authority. The plan reveals Willoughby's commitment to medieval ceremony and reinforces assumptions about Early Modern gender difference through traditional spatial arrangements.

Hardwick Hall, in Derbyshire, was constructed between 1590 and 1597 probably to a plan by Robert Smythson, whose designs for Wollaton and Longleat were probably known to Bess of Hardwick. When construction began, Bess was a widow. Her fourth husband, the Earl of Shrewsbury, died early in 1590 after an estrangement of over three years. He had been a builder as well. His last project, constructed during the late 1570's, was Worksop, known as "the high house in the Midlands," located in Nottinghamshire.[24]

Most writing about Elizabethan architecture emphasizes the prominent role of the patron in the development of the plan and appearance of the house. There is little doubt that Robert Smythson was responsible for the plan of Hardwick Hall. But while the patronage of Thynne at Longleat, Willoughby at Wollaton, and William Cecil at Burghley and Theobalds has been assigned a central role in the architectural innovation of those houses, Bess's role has been overshadowed in favor of the "genius" of Smythson's plan. However, the construction of Hardwick Hall was not a decision made at the end of Bess's life, nor was it without precedent. Although Hardwick is the most famous example of her building, she had been an architectural patron for more than half her life. Chatsworth and Hardwick Old Hall, as well as her involvement in smaller constructions, proves that architecture was a life-long interest—perhaps even a passion.[25] Girouard notes that prominent noble patrons were intimately involved

---

[24] Mark Girouard, *Robert Smythson and the Elizabethan Country House* (New Haven: Yale University Press, 1983), 115.

[25] Sara French, "Women, Space and Power: The Building and use of Hardwick Hall in Elizabethan England" (Ph.D. diss, Binghamton University, 2000), 144-145.

with the construction of their houses[26] and Bess would have been no exception. Summerson speculates that Bess of Hardwick may have been "the prime mover," in the construction of both Worksop and Hardwick.[27]

Bess of Hardwick's management of building projects may be traced back to her second marriage to Sir William Cavendish. Surviving letters include many references to sending workers to various manors; specific directives to level floors or mend windows; and the accounts of her projects include her signature to approve expenditures.[28] This meticulous attention to detail testifies to Bess's intense interest in building throughout her life.

The plan for the new hall was no doubt completed by Robert Smythson. A variant plan is contained in the collection of his papers at the Royal Institute of British Architects; it shows the cross hall arrangement and the general outline of the building. Smythson not only built Bess's tomb, he also worked for her son Charles Cavendish at Bolsover.[29] Another reason for supposing Smythson to be involved has been overlooked. Surely Bess, whose knowledge of the Elizabethan building world was extensive, would have settled for nothing less than the best—and Smythson was unquestionably the most qualified and best-known builder in England. His reputation would have been known to her not least through Shrewsbury but also through Willoughby and Thynne.[30]

Smythson's plan combined Longleat's Palladian simplicity and Wollaton's elimination of the courtyard along with the re-orientation of the great hall ninety degrees. It is set perpendicular to the façade and is immediately

---

[26] Girouard, *Robert Smythson*, 34.

[27] Summerson, *Architecture in Britain,* 64.

[28] Folger Shakespeare Library, Mss. X.d. 428(82) & (83), Cavendish-Talbot Letters and Mss.X.d. 486 (Cavendish Account Book, 1548-1550).

[29] Summerson, *Architecture in Britain*, 59; Girouard, *Robert Smythson*, 144-149.

[30] Sir John Thynne had proposed marriage to Bess in 1565 after the death of her third husband, Sir William St. Loe (Durant, 53); Bess visited Wollaton in 1592 (Durant, 175-176).

visible upon entrance to the house. The oblique and circuitous route left over from medieval houses and repeated at Wollaton has been eliminated.

To an Elizabethan visitor, this difference must have been striking. Hardwick's screen—two columns atop a narrow, chest-high plinth—is a mere pause when entering. Upon entry, the great hall is visually and physically accessible. The great heraldic overmantel is striking: the antlers of its stags, lifted from the crest of Bess's second husband, reach to the ceiling. Nothing impeded the visitor's progress through this room. There was neither a dais nor high table to mark a place of honor for the lord of the manor. Important rooms and people had to found elsewhere.

At the opposite end of the hall from the entrance, the visitor would be directed to the right, through a doorway, to the great staircase. Wide, shallow steps pull the visitor upward, past decorated doorjambs but always toward the upper floors. At the main first floor landing there is nothing to encourage one to stop. Instead, the visitor begins to see light pouring down the stairs from a dramatic opening above. Above, just before the staircase turns to guide the visitor into the high great chamber on the second floor, the ceiling opens to the towers and light from hundreds of tiny windowpanes pours into the house.

This may have been Bess's goal—to pull the visitor away from the lower floors and up to the showcase on the top floor, where magnificent views of the countryside that Bess owned could be had from every room. In the high great chamber, a painted plaster frieze runs around the room just under the ceiling. It depicts various classical scenes, including Diana and her court, a bear hunt, and a personification of Summer. The overmantel bears the royal arms: thus honoring Elizabeth I in her classical guise as Diana as well as through her crest.

Although Elizabeth I never visited Hardwick, it is not unreasonable to presume that Bess had the idea a later monarch might be received there. The close relationship between Bess of Hardwick's granddaughter and James VI of Scotland meant that royalty was already in residence. Arbella Stuart, the daughter of Elizabeth Cavendish, Bess's daughter, and Charles Stuart, brother of James

VI's father, Darnley, was born in 1574. The early death of Arbella's parents left her in Bess's care and her residence at Hardwick was something like imprisonment after a kidnapping plot was revealed in 1592.[31]

It might be argued that Bess's household did not require the same kind of ceremony and thus not the same spatial arrangements as had been seen in earlier houses and at Wollaton, since it did not include a "lord of the manor." However, the construction of Hardwick was undertaken on behalf of Bess's second son, William, on whom it was entailed. William lived with Bess at Hardwick.[32] Thus, Bess not only eliminated the traditional great hall, but made it impossible for her male descendants to use the hall at Hardwick in a traditional way.

The traditional medieval house linked the great hall to the living spaces of its owner. The hall, parlor, and great chamber, were placed in a ceremonial continuum, grouped together opposite to the serving spaces of buttery, pantry, and kitchen. At Wollaton, this spectrum was maintained by placing the hall in the center of the house—eliminating the courtyard in the process—and placing great chambers above and on either side of it. The hall was still a ceremonial space tied to the lord's lifestyle.

Hardwick's hall is different not simply in its re-orientation but also in its relationship to the rest of the traditional spaces within the medieval house. The pantry and buttery are placed on either side of the entrance end of the great hall and the state rooms are placed on the top floor. Rather than tying the hall to the living spaces of the house, Hardwick plants the great hall *between* the serving spaces. If the spectrum of space in the Early Modern country house has a conceptual framework, then Hardwick radically shifts the location of the great hall within that framework: rather than a living space, connected to parlor and great chamber, Hardwick is now part of the serving spaces, cut off from its original location in the medieval ritual of authority and display.

---

[31] Lansdowne Mss Vol. 71, f. 2 (Bess of Hardwick to Lord Burghley, 21 September 1592).

[32] Durant, *Bess of Hardwick*, 186.

The spaces at Hardwick reveal a difference from other houses of the same period and that difference may quite securely be attributed to the gender of its builder. As Alice Friedman states in her article, "Architecture, Authority and The Female Gaze: Planning And Representation In The Early Modern Country House," "The most innovative architectural designs of the period were built for patrons whose needs failed to fit within conventional design typology, whether because of their intellectual interests, their social status, or their peculiar programmatic requirements."[33] Bess's needs certainly were not the same as Willoughby's or Thynne's. Smythson provided a design that reflected Bess's position as the head of her non-traditional household.

Neither Elizabeth I nor Bess of Hardwick set out to subvert gender roles for women, but each made her own contribution to a subversive agenda. According to Breitenberg, Elizabeth I occupied a position of authority that, "despite a broad and powerful discourse that assumed a natural, divinely ordained basis for authority based on (male) gender and status" overset those assumptions.[34] Bess of Hardwick made a distinct difference in the conceptual use of space, creating room for women to re-conceive their approaches to the domestic interior. Despite the fact that the change in the great hall did not immediately catch on, Hardwick's innovation did create a lasting change in the way country houses were laid out in the seventeenth century. Although later houses did include great halls, most notably Montacute in Somerset, Hardwick's alteration of the use and arrangement of the great hall and state rooms sounded the death knell of the great hall's traditional ceremonial and gendered importance. Even if it was constructed in a similar fashion, Hardwick's innovation proved more lasting than the medieval display of lordly authority.

The traditional great hall, functionally superseded long before Hardwick was built, was fundamentally changed with the active input of Bess of Hardwick.

---

[33] Alice Friedman, "Architecture, Authority and the Female Gaze: Planning and Representation in the Early Modern Country House," *Assemblage* 18 (August 1992): 57.

[34] Breitenberg, *Anxious Masculinity,* 1.

It was no longer connected to the rooms most used by the nobility and their households in the fourteenth and fifteenth centuries. It no longer functioned as the symbolic center of authority in the house. The great hall was no longer great, no longer the symbol of the medieval lord's hospitality and authority. The hall no longer contained a dais for enacting lordly authority; the state rooms overtook these functions. At Hardwick, the state rooms were Bess's dais, elevating her above the hall as the supreme authority within her household. As the Early Modern country house evolved, laying bare the ideology of gender roles in the sixteenth century, Bess of Hardwick's contribution was to reveal that ideology and place a woman firmly at the top of the Early Modern country house.

Sara L. French (SUNY-ESF) received her Ph.D. in architectural history from Binghamton University. She currently teaches at both the State University of New York College of Environmental Science & Forestry and Syracuse University. A previous essay, "A Widow Building: Bess of Hardwick at Hardwick Hall," appeared in *Widowhood and Visual Culture in Early Modern Europe* (2003).

"That Awful Throne": Donne, Behn, and the Culture of Dissection

Roberta C. Martin

The early modern period in the West experienced not only extraordinary political and religious change, but also stunning transformations in science and medicine. Philosophies of anatomy and physiology—and the ways women and men thought about their bodies—had begun, by the middle of the seventeenth century, to undergo a gradual, but profound alteration, at least in the minds and epistemologies of those who had some access to the "new" science. A significant element in this drama was the remarkable display that took place in the renaissance and early modern anatomy theatres of Europe and Britain. These "productions" in fact became so popular and "fashionable" that there arose, as Jonathan Sawday points out, a "culture of dissection" that involved a "network of practices, social structures, and rituals surrounding [the] production of fragmented bodies."[1]

Today in Europe and in North America we are participants in (or constructions of) a Western culture in which sophisticated technologies and modern sciences, including medicine, constitute the dominate discourses in a way they did not, even within a "culture of dissection," before the late nineteenth century. In an uncomfortably dictatorial way, modern science often assumes the authority to decide what is and is not knowledge. So too, modern Western medicine and psychiatry has done its share of establishing by fiat "objective truths" about diseases, conditions, or persons that brutally ignore the subjective experience of individuals receiving medical attention. For that reason, when we explore the natural philosophies and medical practices of the medieval and early

[1] Jonathan Sawday, *The Body Emblazoned: Dissection and the Human Body in Renaissance Culture* (New York: Routledge, 1995), 2.

modern period, we have until recently focused on what we see as medical and scientific "progress." This version of history, Mary Lindeman asserts, "postulates a single passage into the modern world leading resolutely from the dark ages of ignorance, superstition, and suffering into a brighter world of knowledge, science, and abundance."[2] In response, "new historians" and "social historians" of medicine, Lindemann argues, should eschew the more simplistic and iatrocentric methods of many older historical accounts and instead emphasize social context, demographics, the history of emotions and the responses of those who sometimes suffer both from disease and from medical practice itself.[3] The social and cultural approaches to medicine are not without their "problems and blind spots," but in general, the new history, allied with a social constructionist perspective on disease, establishes a richer context for exploring medical practices and the bodies and experiences of those who are the objects of these practices.

Especially interesting for explorations of early modern literature and culture are the more tolerant attitudes toward earlier medical disciplines than accounts emphasizing the "march of progress" allowed. In the context of social history and social construction, a mutual accommodation emerges in earlier times between the perception of bodies and the methods and practices that developed to meet the needs of those bodies. In Foucauldian terms, societies, bodies, and medical practices shape one another and determine one another's responses. They are, in other words, effects of one another. This kind of historicization of the body and of medical practice makes visible within medieval and early modern literature and culture a very different corporeal experience, of which writers like William Shakespeare, John Donne, and Aphra Behn made good use, as I hope to demonstrate.

What makes this kind of investigation even more compelling in the early modern period—especially in the seventeenth century—is that changes in

---

[2]Mary Lindemann, *Medicine and Society in Early Modern Europe* (Cambridge, England: Cambridge University Press, 1999), 1.

[3] Ibid., 1-7.

anatomy and medicine provoked new imaginings of the body; inevitably these changes also accompanied (re)constructions of gender and sexuality. This sort of cultural sea change is complex and multi-layered. However, it is safe to say that the increased number of dissections and the anatomists' rather daring challenge to Galenism were important components of the early modern medical transformations that started from the entrenched Hippocratic-Galenic "system" of the late medieval period, were influenced in subtle but important ways by an "empirical" Paracelsian medicine, and emerged with significantly changed understandings of physiology and medical practice with the advent of Gabriel Harvey and the "new" science almost one hundred years after Vesalius.

As Andrew Wear points out, anatomy in the sixteenth century "underwent the same humanist renewal as the rest of medicine."[4] Wear is quick to add that his observation should not be associated with the older historical attitudes to which Lindemann so lucidly opposes a social and cultural approach. Wear therefore first gives careful attention to the superb artist-illustrators to whom the description and precision of anatomical knowledge is so much indebted: Andrea Verrocchio (1435-88), who was Leonardo da Vinci's teacher, Andrea Mantegna (d. 1506), da Vinci himself (1452-1519), Raphael (1483-1521), Albrecht Dürer (1471-1528), and Michelangelo (1475-1564). Important contributions to anatomical method and investigation were also accomplished by the anatomists Gabriele Zerbi (1444-1505), Berengario da Carpi (c. 1460-c. 1530), Realdo Colombo (d. 1559), the earlier, "Hellenistic" anatomist, Alessandro Benedetti (d. 1512) (disliked by both Zerbi and Berengario in part for his rejection of Arab influences on Greek medical knowledge), and Giovanni Batista Canano (1515-79).[5]

Most recent discussions of early modern anatomists and anatomical

---

[4] Andrew Wear, "Medicine in Early Modern Europe, 1500-700," Chapter 6, in Lawrence I. Conrad, et al., *The Western Medical tradition: 800 BC to AD 1800* (Cambridge, England: Cambridge University Press, 1995), 264.

[5] Ibid., 264-73.

explorations emphasize that neither dissection itself nor more accurate representations of the physiology of the human body appeared overnight or in the work of just one or two great anatomists and physicians. Anatomy was not held in high esteem in medieval or early modern medicine, primarily because Galen's anatomical work and insistence on dissection (even though it was animal, not human dissection) was unknown in the medieval period and therefore not a part of the Hippocratic-Galenic system that was well consolidated in the sixteenth century. Until the first part of Galen's *On Anatomical Procedures* was translated into Latin by Guinter von Andernach in 1531, anatomy did not fully begin its radical interventions into medical knowledge.[6] Despite the hold traditional medicine had on both practice and imagination from the fifteenth century onward, physicians and anatomists had begun to question both the methods and the content of Hippocratic-Galenic physiology in the century before Vesalius.

However, no matter how sensitive the social and cultural historians are about avoiding iatrocentrism, none can ignore the centrality of Andreas Vesalius (1514-64), who in fact "helped to restore Galenic anatomy but only to correct it and go beyond it."[7] His "massive" and "brilliantly and profusely illustrated" *De Humani Corporis Fabrica (On the Fabric of the Human Body*, 1543) is, according to almost every source, the most famous anatomy in the Western world and, as Wear observes, "marks a turning point in the medical view of the structure of the body." Vesalius was the first to test Galenic anatomy consistently and methodically by investigating the human body itself.[8]

Important as Vesalius, the *Fabricus*, and the new focus on the anatomist and the anatomy theaters were, however, at this stage I wish to redirect our modern gaze. Center stage in (arguably) the first Western anatomy theater in fourteenth-century Bologna, central to the scenes at the sixteenth-century

[6] Ibid., 270.

[7] Ibid., 273.

[8] Ibid., 275.

university in Padua where Vesalius elevated the reputation of anatomists and anatomy with his demonstrations, and in the foreground of any forum that employed the same sort of autoptic "regard," lies another figure whose visibility we owe to the social and cultural historians: the transgressive and transgressed body itself that becomes the subject of discussion and the object under the knife before a rapt audience of students and citizens—a figure that held even more fascination for its audience if it were the body of a woman.

In fact, it is the opened, empty womb that occupies the exact center of the great title page of the first edition of Vesalius's *Fabrica*. That famous scene teems with the bodies of the living—mostly men, drawn to their version of the frontier to gaze into the black hole of a new space, the interior cave within a female body whose very emptiness reveals another kind of secret as well. If this doomed woman, most likely executed for a crime, "pled the belly" unsuccessfully in an effort to escape punishment, her uterine emptiness reveals her to be a liar, as well as a criminal.[9] There seemed to be no end to the revelations that the anatomists could expose. The price of such revelations, of course, was a peculiarly complete kind of physical dissolution, and the tale persists that some people worried that God might not be able to relocate and reassemble body parts in time for the resurrection. Further, as Sawday discusses at length, the object of autopsy inspired an erotic fascination with, and fear of, that bodily interiority itself.[10]

Sharing stage center with the radically passive and medically violated victim in the Fabrican scene of exploration is the figure of the young Vesalius, who, as a medical student, risked punishment to snatch bodies from the gibbet where they swung in morally salubrious display. His desire was to expose them

---

[9] See the brief account by Jack Pliant, "'Pleading the Belly'—Angela Barnett Avoided 18th Century Death Sentence," (2005), accessed online, February, 2007, *geneologymagizine.com*, http://www.genealogymagazine.com/ plbeanbaav18.html, for some interesting American primary source material on pleading the belly.

[10] Sawday, 9-15.

to another, more intimate and intrusive sort of exhibition. In the theater of the *Fabrica* the naked female victim gazes up at Vesalius, but, of all those in the animated tangle of spectators, he alone turns to meet our gaze, inviting us in to where his hand rests proprietarily on the lip of the open wound, enticing us to become partners in the seductive inquiry of an evolving discipline. From the literary evidence, the metaphors of this neoteric exploration, especially in combination with metaphors of geographical exploration and of sexuality, exerted a powerful hold on the imagination of poets and writers of the seventeenth century.

For this reason the scene will shift eventually to another kind of "execution"—a poetic one in which the anatomy theatre is both the bedroom and the poem, but the fragmented body presented supinely for our investigation is still that of a woman. We will first examine the dissection, exploration and act of possession that John Donne (1572-1631) executes on the body of his lover in "Elegy 19: To his Mistress Going to Bed."[11] Poet, explorer, and brilliant autoptic narcissist, Donne in this exercise wields over the female body the full erotic power that the potent conceits of exploration and anatomical exploitation confer upon him. But the autoptic production does not end there. At the end of the century the clever Aphra Behn (c.1639-1689), successful, if notorious, Restoration poet and playwright, sticks it to the poet-anatomists in her poem "The Disappointment," transforming the violated female body into a weapon that renders impotent the "anatomist's" blade.

First, however, we return to the ideas of ritual fragmentation about which Sawday speaks. The European Renaissance, Sawday argues, reveled in "particularization." Its delight in all kinds of "partition" pervaded "all aspects of social and intellectual life: logic, rhetoric, painting, architecture, philosophy, medicine, as well as poetry, politics, the family and the state....[and] the pattern

---

[11] All quotations from Donne's poetry, unless otherwise specified, are taken from *John Donne's Poetry: Authoritative texts, Criticism,* 2nd ed., Arthur L. Clements (ed.), A Norton Critical Edition (New York: W. W. Norton, 1992).

of all these different forms of division was derived from the human body."[12]

In the context of our discussion of shifting medical paradigms in the early modern period, the importance of the body as an emblem of partition and fragmentation makes dissecting the human body specifically to examine its structure a political act. Thanks to the cultural historians, we can now see two bodies lying on the dissection table instead of one: a humoral body and a body restructured by the newly defined anatomical gaze, even when that gaze recognizes features that the anatomist still identifies as Galenic. Such instantiations of dismemberment served to legitimate the claim that anatomy was the true source of authoritative knowledge about the body. The traditions that these "new" anatomists defied were indeed rich and formidable. The historical process, however, had obscured some elements of the ancient traditions and these lost practices provided precedents for the new perspectives that Berengario, Zerbi, and Vesalius sought to constitute within the eyes and understanding of their beholders.

Although anatomists certainly disagreed about the nature of the body, medicine, and the natural philosophy that was so important to the Galenic tradition, they opposed dominant medical views that (contrary to the ones that Galen himself actually held) devalued anatomy and dissection as sources of knowledge about the body, health, disease, and healing. In reality, the various paradigms of medical knowledge that informed perceptions of the humoral body reflected, in varying degrees, ancient Greek, Roman, and Arabic methods of healing. These methods stretched back not only to the dominant humoral and philosophical traditions of Hippocrates (450-370 B.C.E.) and Galen (129-200\216 C.E.), but also to Galen's Greek Alexandrian predecessor and rival theorist, Erasistratus of Ceos (ca. 304-ca. 250 B.C.E), who continued the anatomical and autoptic investigations of Herophilus (fl. 280 B.C.E), also of Alexandria, where human dissection—and perhaps even human vivisection—were practiced. Even

[12] Sawday, 3.

though the earlier medieval period during which dominant physiological paradigms were constituted was not aware of it, Galen was also vitally interested in anatomy and was skilled in dissection (although he practiced animal rather than human dissections). Galen differed from Erasistratus in his elaboration of the theory of humors and in a version of natural philosophy that emphasized unified teleological and theistic approaches to anatomy and physiology.[13]

Despite the fact that ancient medical practitioners had stressed anatomical knowledge, it was an extremely pliant and adaptable Hippocratic-Galenic tradition insisting on the fluid nature of the body that survived into the medieval period and that devalued organic and anatomical knowledge. This version of the various ancient medical traditions emphasized the Galenic notion of the four basic humors—phlegm, blood, black bile and yellow or red bile—as well as the "qualities" (e.g., hot and cold, moist and dry). These notions and these perceptions of the life of the body were what the medieval period inherited as "Galenism." Vestiges of this medical tradition and practice lasted in one form or another almost until the end of the early modern period.

As Mary Lindemann points out, the "hardy construct that was 'Galenism' needs to be distinguished from Galen the man," not least because the medieval period did not have that prolific writer's complete works.[14] Two of the most important missing texts were *on the Use of the Parts*, as noted above, and *Anatomical Procedures*.[15] Moreover, all of what the medieval West knew as "Galenism" had detoured through Arabic sources. However, despite the inevitable changes that come when any "system" passes through different ages and cultures, the physiology and its attendant images and attitudes about the body that the early moderns had inherited from medieval times could still be called

[13] Vivian Nutton, "Medicine in the Roman World, 250 BC to AD 200," Chapter 2 in *The Western Medical tradition: 800 BC to AD 1800* (Cambridge, England: Cambridge University Press, 1995), 65-67. Also see Lindemann's discussion of Galen in *Medicine and Society*, 67-70.

[14] Lindemann, 68.

[15] Ibid.

"Galenic" to a degree. Also, as Lindemann suggests, learned and lay medicine overlapped considerably, so standard "Galenic" physiological concepts and perceptions were widespread.[16]

Also important in the Western medical tradition were Aristotle's practical and philosophical notions, including his natural philosophy, which differed from Galen's highly developed, theistic version in that "designing nature," although immanent and active, was also impersonal.[17] Interestingly, Aristotle (384-322 B.C.E) may have ascribed a totally different theory of the humors to the historical Hippocrates; the theory that by the second century was firmly associated with Hippocrates may have been derived from Polybus (fl. ca. 410 B.C.E), a disciple and son-in-law of Hippocrates. The importance of this ascription, or more precisely the possibly mistaken ascription, to Hippocrates is that it indicates that the recognized humoral tradition of the early modern period was, back in the day, simply one theory among many. As Vivian Nutton observes,

> The theory of the four humors is only one of a variety of attempts in early Greece to explain health and disease in terms of bodily fluids, humors. For centuries there was no canonical number of humors, nor any firm agreement as to which fluids were to be placed in that number.[18]

Also lurking in the background of the complex traditional history of medicine were other groups of healers and practitioners like the ancient Roman Methodists, Empiricists, Erasistrateans and Pneumatists as well as physicians like Tiberius Claudius Menecrates, who founded his own medical sect. In the medieval period, "Galenic" medicine also owed its rediscovery and survival to the

---

[16] Ibid.

[17] Nancy G. Siraisi, "Vesalius and the Reading of Galen's Teleology," *Renaissance Quarterly*, 50:1 (Spring 1997): 4.

[18] Vivian Nutton, "The Fatal Embrace: Galen and the History of Ancient Medicine," *Science in Context* 18:1 (2005): 116; *Islamic Culture and the Medical Arts: Late Medieval and Early Modern Medicine*, http://www.nlm.nih.gov/exhibition/islamic_medical/islamic_14.html.

great Islamic translators and physicians like Rhazes (Abu Bakr Muhammad ibn Zakariyya al-Razi, c. 850-925), Avicenna (Abu Ali al-Husain ibn Abdallah ibn Sina, 980-1037) and Averroes (Abu'l Walid Muhammad ibn Rushd al-Qurtubi, 1126-1198).[19]

Clearly, by the early modern period—a time we tend to think of as the "real" beginning of anatomical investigation and medical practice—a rich, varied, and complex history of corporeal perception and physiological understanding had conditioned the lived experience of men and women since before the common era. Against this teeming background of humoral perception, the sixteenth century anatomists stepped into the anatomical theatre to challenge the humoral gaze and revise the anatomical imagination.

There is, however, no denying the probable excitement, drama, horror and trepidation experienced by those who went to peer for the first time into the recesses of a body most likely brought fresh from the gallows—a body that, mother, merchant or murderer, would, they presumed, reveal an interior very much like their own. But what did they expect to see?

According to Giovanna Ferrari,

> The teaching of human anatomy through dissection is unanimously thought to originate in Bologna...at the beginning the fourteenth century, around the figure of Mondino, in response to the requirements of university life and to the need to train future doctors and surgeons.[20]

---

[19] Ibid., 115. For discussions of ancient Greek and Roman medicine and medicine in late antiquity, see Nutton, Chapters 1, 2, and 3 in Conrad, *The Western Medical Tradition: 800 BC to AD 1800*. The authors of this text are members of the Academic Unit of the Wellcome Institute for the History of Medicine, London. See also Lindemann, *Medicine and Society*; Edward Grant, "When did Modern Science Begin," *American Scholar*, 66:1 (Winter 1997): 105-114; Roger French, "Parts 1 and 2," *Medicine Before Science: The Business of Medicine from the Middle Ages to the Enlightenment* (Cambridge, England: Cambridge University Press, 2003). For a discussion of the influence of Arab medicine on Medieval Europe, see Lawrence I. Conrad, "The Arab-Islamic Medical Tradition," Chapter 4 in *The Western Medical Tradition*; Charles F. Horne, "Introduction," vol. 6: *Medieval Arabia in The Sacred Books and Early Literature of the East* (Parke, Austin, & Lipscomb, 1917): 90-91, http://www.fordham.edu/halsall/source/1020Avicenna-Medicine.html, accessed online, February, 2007, online Medieval Sourcebook, http://www.fordham.edu/halsall/sbook.html.

[20] Giovanna Ferrari, "Public Anatomy Lessons and the Carnival: The Anatomy Theatre of Bologna," *Past and Present*, 117 (November, 1987): 5. Although I have no doubt that this

We can with certainty assert that the fourteenth century individuals who attended the anatomy theatre in Bologna went there, experientially, in a humoral body, looked out from a humoral body, and expected to see opened up before their eyes a body that was also humoral. Conversely, just before the mid-seventeenth century, although the humoral body was still alive and well in many respects and in most places, "the anatomy lesson" had become so much a part of early modern thinking and rethinking that there arose "anatomies" of all kinds.

As noted above, the practice of division and partition invaded nearly every aspect of early modern living, as Robert Burton's *Anatomy of Melancholy* and John Donne's "The First Anniversary: An Anatomy of the World" attest. It is therefore not surprising that newer images of the body and of body parts informed by the so-called "new science" invaded the poetry and other literatures of the early modern period. These beginnings of a new imagination competed with the older, traditional, fantasies of the body that reflected what seems to us the rather peculiar quotidian experience of inhabiting the mind and skin of the fourteenth century and the seventeenth century woman or man.

However, before we speculate about what autoptic exposure of hidden corporeal recesses may have been disclosed to the medieval and the early modern person, I wish to consider for a moment just one aspect of the kind of body through which most modern individuals (or post modern, depending on one's theoretical inclinations) experience their interior physical life. In fact, we can do it by heart.

A brief glance at the way the blood moves through the body reveals at once the difference between our contemporary imaginative perception of our corporeal existence and that of the medieval or early modern individual. To recognize this difference helps us to understand how our own embodied

---

statement is true for the medieval and early modern period, I wonder, if, given what we know—or rather don't know—about ancient anatomical practices, especially in Alexandria, we might qualify Ferrari's statement slightly and leave open the possibility that in the ancient world Erasistratus and others, especially in Egypt, may have taught anatomy through the dissection of human cadavers.

experience can affect our ability to "read" embodied experience in the poetry of another age. As readers, we are situated within (or perhaps "subjected to") bodies that we visualize as possessing an "'anatomy of solid parts'...dependent on the "'physiological interplay of organs,'" as Barbara Duden calls them.[21]

Today, if we contemplate our own interior spaces, we most likely imagine organs first, the heart pumping blood through the lungs where erythrocytes seize oxygen. The organic pump then drives that highly reactive gas through a complex web of arteries to the organs of the digestive system for supplies: last night's dinner is mechanically and chemically broken down into nutrients and absorbed into the blood and lymph vessels through the intestinal walls. Steadily the body's version of a perpetual-motion machine propels the blood through the liver and out to nurture every cell in every nerve and muscle and organ, only to be thrust back through the venal system to begin its circulatory delivery rounds once more.[22] This one organic action is a small but important part of the physical experience that forms and informs our sense of our bodies and our relationship to the world.

We should consider one more detail about our modern corporeal experience—the importance that boundaries have for us—and in a different sense, had for those who preceded us. We see and feel our skins and the skins of others as impermeable, protective barriers between the tender mechanisms of our bodies and the outside world. What's outside the body pretty much stays outside unless we invite it in, and, when we are healthy, what's inside egresses in limited amounts on a relatively regular and controllable schedule. As readers, we tend to visualize the activities of the early modern bodies written about by poets and the bodies of the writers in this way, and if we do, we unwittingly participate in

---

[21] Barbara Duden, "Kapitel aus der Säftelehre," quoted in Ulinka Rublack, "Fluxes: the Early Modern Body and the Emotions," trans. Pamela Selwyn, *History Workshop Journal* 53 (2002): 2.

[22] Kent M. Van De Graaff, *Human Anatomy* (Columbus, Ohio: McGraw-Hill, 2001), 538-71.

physiological anachronism and further a kind of cultural illiteracy. By virtue of perception and imagination, the men and women of the seventeenth century lived in a different body and therefore a different world. Moreover, I wish to argue, because of those different bodies, not just their genders but also their sexes and their sexual bodies were perceived as quite different from our own.

First, the early modern body was not a self-contained and enclosed entity, but, in Bakhtin's famous description, it "was constantly changing, absorbing and excreting, flowing, sweating, being bled, cupped and purged."[23] Influences and vapors flowed in at will, altering the health and attitudes of the body; humors, juices, and, in extremis, vital spirits flowed out. Bodily organs were critically important, but only in relation to the fluids upon which they depended. The heart sank or trembled, and blood "froze," not metaphorically, but literally. Ulinka Rublack cites the case of an Ambassador to Maximilian II who reports that on a trip, he stopped twice because "[his] blood began to flow from [him] in large quantities along with urine, but without any sensation of pain or greater inconvenience of any sort."[24]

If such a physiological event happens to us, we rush to the emergency room, but for the ambassador, this flow was thought to be beneficial, an adjustment of his vital juices, a rebalancing of the fluids to which the early modern person was so attentive. Rublack notes that many symptoms described by early modern sources are quite alien to our own systems of meaning, and she asks the fascinating question, could it have been that in this period, "not only bodily existence was conceived of differently from today, but also that this body actually behaved differently?" Rublack observes that other scholars have asked this question as well.[25] Therefore, not only the historical, but even the ontological status of the body may vary with the epistemology that organizes it. This

---

[23] Michail Bakhtin, *Rabelais and his World*, quoted in Rublack, "Fluxes," 2.

[24] Ibid., 1.

[25] Ibid., 2.

possibility bears directly on the subject of the sexing of bodies, as do the four humors in an even more important way. These mobile substances—black bile, yellow (or red) bile, blood, and phlegm—constituted the physiology and psychology of the early modern individual. If the humors were unbalanced by inner or environmental forces, then illness occurred. Diseases were specific to the individual alone; specific disease entities like plague or tuberculosis, as we think of them, did not exist because general categories of disease did not exist.

As we have seen, from the time of Hippocrates and Galen until the mid to late eighteenth century, a mixture of what Lindemann calls "environmentalism and humoralism" conditioned perceptions of the body and its health.[26] The emotions, too, not only were tied to the physical condition of the body—to the balance of humors—but also were thought to "flow." Rublack observes that the classical humanist education taught that the proper balance of emotion necessitated both the "relativizing and the expression of feelings." Like Freud three centuries later, Francis Bacon (1561-1626) felt that repressed emotion was bad for one. In fact, "the chief benefit of friendship was 'the ease and discharge of the fullness and swellings of the heart,' which was the root of all passions." Blockages in the body were dangerous, and so, too, were those in the mind. One took medicines "against congestion of the liver, lung or brain, 'but no receipt openth the heart but a true friend," to whom one confided "his" feelings (since only men were the intellectual/emotional equals of other men, a man's "true friend" was almost always another man). Mind and body were interrelated. If feelings "swelled the heart," then such swellings were physically and mentally dangerous unless relieved by "verbal transmission."[27]

Disease and health, both physical and emotional, were a precarious continuum dependent on the balance of humors and other bodily fluids which, moreover, were "fungible": that is, they were interchangeable and capable of

---

[26] Lindemann, 9.

[27] Rublack, 3.

replacing one another within the body. In Act I, Scene V of Shakespeare's *Macbeth,* Lady Macbeth invokes the "spirits who tend on mortal thoughts" to perform a series of physical and fungible transformations within her fluid body:

> [U]nsex me here
> And fill me, from the crown to the toe, top-full
> Of direst cruelty! make thick my blood,
> Stop up th'access and passage to remorse;
> ....Come to my woman's breasts,
> And take my milk for gall, you murdering ministers.
> (*Macbeth*, 1.5.41-43, 47-48).[28]

Lady Macbeth is not speaking as metaphorically here as we are used to thinking. Since sex, too, depended for the most part on the humors—men were hot and dry and women were cool and moist—by heating up the "lady" with the "smoke of hell," thickening her blood so it could not "flow," creating a blockage of sympathy by stopping up the passages of the body so that foul humors could not escape and literally transforming her fungible milk to gall, the spirits are asked to construct of her, neither a man nor a woman, but a demonic figure of raging ambition produced by physical (and therefore emotional) imbalance.

It is also worth pointing out, as Patricia Crawford does, that sexuality and health during the early modern period were linked. In fact, medical theorists believed men and women needed sexual activity in order to maintain good health. Seed accumulated in the bodies of both males and females, but because men were hotter they "concocted" more seed and therefore their urges and needs were more powerful. Starting in adolescence, the healthy body had to discharge or reduce the amount produced. If marriage, which, according to religious discourse, was "the only site of the lawful sexual activity that was necessary for the health of both sexes," was not possible, or if a partner were unavailable, then physicians recommended sober diet and vigorous exercise: overeating caused a person to be

[28] William Shakespeare, *Macbeth*, ed. Kenneth Muir, Arden Edition of the *Works of William Shakespeare*, 35 vols., general ed. Richard Proudfoot, (1984; reprint, New York: Methuen, 1986), 30.

full of the "spirit of fornication." Since men produced more seed, their sexual urges were more powerful and they were "less able to be chaste."[29]

The poets John Donne, Andrew Marvell, and George Herbert in the first half of the seventeenth century constructed elaborate and often erotic metaphysical conceits from the imagined and imaginary materials of their humoral bodies. At the end of the century, after William Harvey had demonstrated that the heart is indeed a pump (a fact of which European anatomists were unaware until Harvey's demonstration) and that the venal and arterial systems operate in a single continuous loop instead of being separate systems,[30] Aphra Behn made sly use of the "old" Galenic humoral traditions to play daring poetic games that destabilized notions of sexuality and gender. Through the complexities of early modern physiology and through the bodies of these poets—the bodies presented in the poems and the bodies, in a culturally collective sense, from which the poets themselves project their various and variable personas—we can discover remarkable interpretive possibilities and new ways of perceiving, not only bodies, but the sex and gender of those bodies as well.

Donne, as we well know, was a voracious consumer of metaphorical

---

[29] Patricia, Crawford, "Sexual Knowledge in England, 1500-1700," in *Sexual Knowledge, Sexual Science: the History of Attitudes toward Sexuality,* eds. Roy Porter and Mikulas Teich, (Cambridge, Mass.: University of Cambridge Press, 1994), 85.

[30] William Harvey, The *Anatomical Exercises: De Motu Cordis* and *De Cirulatione Sanguinis in English Translation*, ed. Geoffrey Keynes (London: Nonesuch Press, 1953 under the title *The Anatomical Exercises of Dr. William Harvey, De Motu Cordis* 1628; *De Circulatione Sanguinis,* 1649; reprint, New York: Dover, 1995). The first English translation of the two works was published by Richard Lowndes, London in 1653. The original Latin text of *De Motu Cordis* was published in 1628 and the Latin text of *De Circulatione Sanguinis* was published in 1649. It is interesting to note that our sanguine Eurocentrism has kept us from acknowledging—or even knowing—that prior to Harvey's discovery of the circulation system, a celebrated Muslim Syrian doctor, Abu-Alhassan Alauldin Ali Bin Abi-Hazem Al-Quarashi, known as Ibn Nafis, born in Damascus ca. 1210, discovered the pulmonary circulation of the blood and knew that the heart was a pump in the 13th century, three centuries before Servetus, Vesalius, Colombo, and then Harvey worked it all out. See Ayman O. Soubani, and Faroque A. Khan, "The Discovery of the Pulmonary Circulation Revisited," The New England Medical Center, Tufts University School of Medicine (Dr. Soubani), and Boston and the Department of Medicine (Dr. Khan), Nassau County Medical Center, SUNY at Stony Brook, East Meadow, New York, 1995, accessed on line January, 2007, http://www.kfshrc.edu.sa/annals/152/mh9422ar.html.

material of all kinds—alchemical, astronomical, geographical, philosophical, and religious, but anatomical metaphors seem peculiarly appropriate to his hypervigilant, if not hypochondriacal, sensibilities. Disease, the secrets of the body, and the new knowledge produced in anatomy theatres held both fear and a fascination for him, as these lines from the "Second Anniversary" attest:

> Have not all soules thought
> For many ages, that our body'is wrought
> Of Ayre, and Fire, and other Elements?
> And now they thinke of new ingredients.
> . . . . . .
> Knowst thou but how the stone doth enter in
> The bladders cave, and never breake the skinne?
> Know'st thou how blood, which to the heart doth flow,
> doth from one ventricle to th'other goe?
> (263-66, 269-72)[31]

There were mysteries hidden even in dissected bodies, and they were not comforting ones.

In Galenic theory blood traveled to the right side of the heart through invisible pores or tiny holes in the septum of the left side of the heart. There it mixed with air and "spirit" before traveling to the rest of the body. The venous system was completely separate from the arterial circulation except when the blood passed through the tiny holes in the septum. Donne, who could be famously gloomy about the revelations of the new anatomical "science," bemoans the loss of the ancient elements—earth, air, fire, and water—that in Hippocratic-Galenic physiology constituted the humors and provided an important link between microcosm and macrocosm.

Donne of course doesn't know that the heart is a pump, but he does seem to know that there is a problem with the Galenic theory of cardiac pores. Moreover, unlike most people who accepted that the skin was porous and that fluids and "influences" of many kinds flowed through it in both directions, he

---

[31]John Donne, "Of the Progresse of the soule: the second Anniversary" in *John Donne: Poetical Works,* ed. Sir Herbert Grierson (London: Oxford University Press, 1933), 234-35.

seems to see the skin as a barrier and is puzzled over "stones" in the "bladder." In this passage, Donne is disturbed by the contradictions between old traditions and new science. He is losing his confidence in ancient phenomena without having adequate explanations of the body to replace them.

Nevertheless, Donne understood the stunning potential inherent in a convergence of geography and physiology. On more than one occasion, his own body becomes a chart or a kind of "anthropomorphic map" whereupon he traces the courses of fevers in both body and soul:[32] "Whilst my physicians by their love are grown/ Cosmographers, and I their map, who lie/ Flat on this bed, that by them may be shown/ That this is my South-west discovery....As West and East/ In all flat maps (and I am one) are one,/ So death doth touch the resurrection" (128.6-7, 14-15). In the lines that occur between these that I have quoted from "Hymn to God my God, in my Sickness," Donne indulges in a bit of directional dissection, yet the effect of the metaphysical union of geography and physiology is to "discover" in his own anatomy a powerful unity. West and East, death and life, body and soul become one, and through the process of the poem, Donne resurrects himself intact.

What Donne does to the body of a woman, however, is precisely the opposite of "intact" and constitutes an autoptic masterpiece. I suggest that conceits of geographical discovery and metaphors of dissection were not only omnipresent in the seventeenth century imagination, but that they were universally and powerfully connected as two dimensions of the same exploratory urge, especially for Donne. For this reason, metaphors of dissection and dismemberment resonate within the language of disrobing and exploration.

In "Elegy 19: To His Mistress Going to Bed," the impatient Donne first approaches his mistress like a soldier—or explorer—anxious for battle: "The foe oft-times having the foe in sight/ Is tired with standing though he never fight" (3-

---

[32] For an interesting discussion of the so-called "anthropomorphic map tradition, see Claude Gandelman, "The Poem as Map: John Donne and the 'Anthropo-morphic Landscape' Tradition," *Applied Semiotics / Sémiotique appliquée* 3:8 (1999): 207-215; (Arcadia: *Zeitschrift für VergleichendeLiteraturwissenschaft,* 19 (1984): 244-51.

4). This is a seduction poem, however, and the knife that will strip away layers of resistance, the way the anatomist strips away layers of flesh, is the metaphysical conceit. Of course, in these lines there is also another, fleshier, penetrative instrument still "standing" that the poet wishes to wield as well.

"Off with that girdle," Donne demands, because it "encompasses" a "world" fairer than heaven that the poet wishes to explore. The body of his mistress has become a continent protected by her clothing in the way the body's interior is protected by layers of flesh. Next Donne urges off the "spangled breastplate" and the "happy busk," or corset, both evocative of the body parts that are revealed. The "wiry coronet" goes next, revealing a "hairy diadem" which is easily a displacement upward of the erotics of other hairy places. When these body elements are partitioned and exposed, it is time to shift the action to the center of the anatomy theatre: "Now off with those shoes, and then safely tread/ In this love's hallowed temple, this soft bed (17-18). Displacement dissolves as Donne's "Angel brings with [her]/ A heaven like Mahomet's Paradise" (20-21) that "sets his flesh upright" once again (24).

Like the true, well-equipped explorer that he is, Donne asks for a charter to discover the secrets of "his" continental body:

> License my roving hands, and let them go
> Before, behind, between, above, below.
> O my America! My new-found-land,
> My kingdom, safeliest when with one man manned....
> To enter in these bonds is to be free;
> Then where my hand is set, my seal shall be.
>
> (26-29, 31-32)

From a whole woman protected from intrusion by a "skin" of breast-plates and busks, Donne's mistress is now a collection of exposed body parts transmogrified into a piece of property not her own, but "manned" and "mined" by the triumphant "free" poet and subdivided into parts—those parts "before, behind, between, above," and "below—" with each parcel "sealed" away, presumably, from any other anatomist-explorer who might desire to "rove" a little, sword at the ready. Moreover, Donne places his hand on his supine and passive mistress close to the spot where, on the title page of the *Fabrica*, the proprietary and well-clothed Vesalius, another triumphant explorer, places his on the naked body of the woman that he has "unsealed" and exposed to an even greater degree than the seducer Donne.

Considered from this last perspective, the final lines of this ebullient hymn to exploratory dismemberment seem a bit disingenuous: "to teach thee, I am naked first: why than,/ What needst thou have more covering than a man?" (47-8) The "nakedness" here is hardly equal and scarcely equally unified. The "whole" poet has not been anatomized, and his body, except for his triumphantly raised "blade," is still draped in poetic silence as he coaxes his mistress out of the last of her linen in order to be pinned down and explored in earnest. She is fixed, fragmented, and revealed; he is not.

Do we want poetic revenge? We do. But we have to wait until the end of the century to get it, when Aphra Behn (c.1639-1689), who was at one time the most popular playwright in London, takes the poetic stage instead—a woman who is not afraid to wield a few metaphors of her own. A successful Restoration poet, playwright, novelist, spy, and libertine, Behn's major literary activity occurred precisely during the period when notions of gender were most in flux. Between 1670 and 1678 Behn wrote some seventeen plays, including *The Rover*, and later produced a large body of poetry. She also (arguably) pioneered the novel form in England with *Oroonoko* (1688) and *Love Letters between a Nobleman and his Sister* (1684-87). As the first professional woman writer in England to support herself entirely through her writing over a long period of time, she defied the customs and attitudes of the day by seeking acknowledgment of her literary talents and by demanding payment for her literary efforts. This adventurous defiance of social orthodoxy found expression in her literary works and, I submit, is striking not only in her subversions of gender, but especially in her manipulation of sexual categories.

But what about sexes? If the bodies were unstable, changeable, and subject to imbalance, what made the division of the sexes so clearly delimited and defined? The answer, of course, is that they weren't. Thomas Laqueur argues that until the mid 1750's people recognized a "one-sex" body. The female body was simply an inferior version of the male body; female parts were for a long time not separately named, and the vagina was considered simply an undescended penis. In this world there were at least two genders, but one sex—the male one.

The boundaries between male and female were ones of degree and not kind. In this perceptual and cultural environment—and in its texts—"*sex*, or the body, must be understood as the epiphenomenon; *gender*, what we would take to be a cultural category was primary or 'real.'"[33] In other words, one's organs mattered less than one's clothes for determining sex and at least one account persists that a woman who raised her body heat too radically or engaged in "violent" exercise caused her internal organs to "pop out," thus transforming her into a man. From what we know of the humoral body, this account of gender and sex is quite plausible, even if it is perhaps in need of some complication.

Laqueur's concept of the "one body" model of sexual difference, in addition to altering the trajectory of much of the ensuing research in both sexual history and the broader social history of the early modern period, also generated an important debate over his use of sources and over issues of body and sexuality in general. For example, significant objections to Laqueur's timeline have been offered by Robert Martensen and Mary Fissell, who argue for an earlier, seventeenth century, understanding of the "two body" model. During the ensuing conversation among scholars, some of which is outlined by Tim Hitchcock in *English Sexualities, 1700-1800*, Lorraine Daston, Katherine Parks, and Gail Kerns Paster, among others, have shown that this model, while it may have been prominent in medical understanding and social practices, in fact co-existed with the two-sex model from the ancient Greeks until well into the 18th century, when

33 Thomas Laqueur, *Making Sex: Body and Gender from the Greeks to Freud* (Cambridge, MA: Harvard University Press, 1990), 8; Tim Hitchcock, *English Sexualities, 1700-1800* (New York: St. Martin's Press, 1997), 45-47.

our present bi-polar model of gender and sexuality became fully entrenched.

Nevertheless, the idea of stable binaries of sex and gender in the early modern period and before becomes less tenable as we comprehend that genitals were not the primary markers of sexual category. The fluid, unstable bodies perceived by early modern cultures were quite likely the sites of fluid, unstable sexes, as well. Michael McKeon argues that until the eighteenth century, the "one sex" that existed "was a sociological rather than an ontological category":

> Men and women exist on a continuum whose basic discriminants are social rank, cultural role, and legal entitlements, not organic identity. Embodied sexuality was relatively elastic and fluid. "Nature" was not conceived as a physiological bedrock stabilizing sexual personality; and the distinction between the biologically grounded category "sex" and the socially constructed category "gender" was therefore largely unintelligible.[34]

Only in the late seventeenth century in England, according to McKeon, was there the "recognition" of a "natural," or biological division between sexes believed to be distinct and therefore incommensurable. As McKeon argues, only then did the sexes become defined "not by behavior, which is variable, but by nature, which is not." (McKeon here refers, I think, to the more general and ordinary understanding that people had about sexual categories and not to the more scholarly evidence that the two-body theory co-existed in medical literature with the one-body theory all along.) Also at the end of the seventeenth century, during the Restoration, Aphra Behn offers literary evidence that sexual category was indeed still fluid and that traditional humoral notions could still be employed to undermine these categories further.

Behn wrote an "imperfect enjoyment" poem, "The Disappointment,"[35] as part of a group competition that included her pal, John Wilmot, Earl of Rochester

---

[34] Michael McKeon, "Historicizing Patriarchy: *The* Emergence of Gender Difference in England, 1660-1760," *Eighteenth-Century Studies* 28: 3 (1995): 301.

[35] All references to Behn's poetry are taken from *The Works of Aphra Behn*, vol. 1: *Poetry-*, 7 vols. ed. Janet Todd (Columbus: Ohio State University Press, 1992-96).

and other libertine poets. Todd has observed that "Behn became…one of the boys, using male language with a freedom no proper lady could have allowed herself."[36] I have argued, however, that it is not simply a matter of propriety—that to act like "one of the boys" during this period is in a sense to *be* one of the boys. I also think it quite likely that Behn was in some way aware of the mutable sexual categories of her day and deliberately exploited that mutability in order to assume in her poetry subject positions that were "not-male" and "not-female." Elsewhere I have argued that in her poem "To the fair Clarinda, who made Love to me, imagin'd more than Woman," Behn uses the figure of the hermaphrodite to destabilize sexual categories and to assume a dazzling series of not-male, not-female subject positions. In "The Disappointment" she capitalizes on one of the worst libertine fears. Using the traditional characteristics of males and females as outlined in humoral theory, Behn reverses the genders of the participants in a slightly disturbing tongue-in-cheek date-rape scene.

"The Disappointment," however, was published in 1680. Is this in fact too late to be able to employ a convincing humoral context for her manipulations of sexual category? By no means. It is true that William Harvey's definitive demonstration of the pulmonary circulation of the blood in 1628 confirmed the effectiveness of experimental science and eventually would change radically the way scientists, doctors, and ordinary men and women thought about anatomy and physiology, as well as how they experienced their bodies. This new orientation to the body, however, was slow to effect a thorough perceptual or imaginative transformation, either in the way men and women envisioned themselves or in the way most physicians and other practitioners practiced medicine. Moreover, in England, co-existing with Galenic humoral medicine and the empirically based "new science" throughout the seventeenth century was a hermetically based, "empirical" medicine founded primarily on the teachings of Paracelsus (1493-1541). These three medical disciplines differed in their notions about how the

---

[36] Janet Todd, *The Secret Life of Aphra Behn* (New Brunswick: Rutgers University Press, 1997), 194.

body worked, and, to a certain extent, their notions about anatomy itself. Well into the Restoration and beyond, physicians and thinkers struggled to achieve a philosophical framework to change or replace Aristotelian or Galenic natural philosophy and to accommodate the ultimately irrefutable insights of thinkers like Newton, Bacon and Harvey.

Poets like Shakespeare, Donne, and Andrew Marvell mined their metaphors from whatever model of the body suited their purposes. Yet even then, it is likely that Donne's continental conquest, the beautiful young man and the less beautiful mistress of Shakespeare's sonnets, and even Marvell's continentally gigantic "coy mistress" were all leaky bags of fluids, continually exchanging influences with the outside environment. Donne's amatory conquest, his "America! [his] new found-land," was more like a permeable bog, a cool, moist swamp. And who better to take advantage of the humoral nature of bodies and the fluid characteristics of the sexes than Aphra Behn?

In "The Disappointment," Behn tells the story of another sort of anatomical "performance" (a failed one), in which the bodies of the two would-be lovers are definitely humoral. Our anatomy theatre has become a thicket where the shepherd, Lysander, "surprises" the "fair Cloris" and attempts an anatomy lesson of his own (3).[37] Cloris "Permits his Force," but also "protests" by breathing "Cease, Cease" faintly into his ear. Lysander doesn't, of course, and Cloris faints ambiguously away. Unfortunately, Lysander is so overwhelmed by the sight of her body through her "loose thin *Robes*" that he is "too transported" to rise to the occasion.

As we might remember, according to the Galenic theory still current in the seventeenth century, the main difference between men and women lies not in their private parts, but in their body heat and moisture content. Men are hot and dry; women are cool and moist. Lysander's "burning hand" grabs Cloris's "snowy breast"—cool as well as white; Her lips are "balmy," her eyes cast a

[37] Aphra Behn, *Works*, Vol. 1: *Poetry*, no. 28 "The Disappointment," 65-69.

"Humid Light," and love's "altar" is a "Fountain where Delight still flows" (45-59). The overheated and humorally unbalanced Lysander is "o'er-Ravished" with Love's "Sacred Flame," and it is only when he is "unable to perform the Sacrifice" that he is "unmanned" and his ardor becomes decidedly damp—his "Insensible," despite all his frantic efforts, falls "weeping" in his Hand":

> Now all the brisk and active Fire
> That should the Nobler Part inflame,
> Serv'd to increase his Rage and Shame,
> And left no Spark for New Desire.
> (95-98)

Cloris meanwhile returns from the convenient "Trance/ Which Love and soft Desire had bred" and, reaching for "that Fabulous *Priapas,/* That Potent God," finds "Priapas" much limper and cooler than she expected. Like a "shepherdess" who finds "beneath the verdant Leaves a snake," Cloris snatches her hand away, "finding that God of her Desires/ Disarm'ed of all his Awful Fires,/ And Cold as Flow'rs bath'd in the Morning-Dew" (111-13). It seems that our lovers have switched sexes. Not only is Lysander cold and damp, but Cloris suddenly heats up with "confusion": "The Blood forsook the hinder Place,/ And Strewed with Blushes all her Face,/ Which both Disdain and Shame exprest." (116-118) Cloris races off, leaving the emasculated Lysander "fainting on the Gloomy Bed," left alone to curse "the *Shepherdess's* Charms,/ Whose soft bewitching Influence/ Had Damn'd him to the Hell of Impotence": (139-40) despite his "silent Griefs" that "swell up to Storms," he is cold, limp, damp, and faint—in short, he is female.

"The Disappointment" is Behn's response to a fairly long tradition of male impotence poems of which Ovid's poem on this subject is one of the first. Rochester's contemporary "performance poem," "The Imperfect Enjoyment," appears with Behn's version in his own *Poems on Several Occasions*, published in 1680.[38] Behn's immediate source is a poem by the French author de Cantenac,

---

[38] Todd, "Head Note no. 28," in *Works*, vol. 1: *Poetry,* 392-3.

but Behn has made it her own, giving us a good deal of Cloris's view of the matter—a perspective male poets neglected. Obviously Behn is glossing old tales of male vulnerability with an irreverent sense of fun. But I would argue that Behn also comments on another feature of male seduction poems that has been with us at least since Petrarch, a feature that is given additional clarity by Sawday's earlier observations: that tendency of poetic seducers like Donne to build erotic tension in their poetry by anatomizing and dismembering their female subjects.

The Donne of "Elegy 19" is impassioned and empowered by the systematic exposure, partition, and expropriation of his mistress's body. In "The Disappointment," Behn turns the anatomy tables on the seducer by first dissecting and then reassembling the parts of Cloris's body into a whole woman so beautiful that Lysander the anatomist-explorer is transfixed and then dispatched by his own sword. He is first "invigorated" by her "Bright Eyes, sweet, and yet severe"; he kisses her Mouth, Her Neck, her Hair; he presses his trembling hand to her breast and continues his anatomical investigation until, like Donne's soldier-explorer, he discovers America: "All her Unguarded Beauties lie/ The spoils and Trophies of the Enemy" (39-40). But the best is yet to come:

> By swift degrees advancing—where
> His daring Hand that Altar seized,
> Where Gods of Love do sacrifice.
> That Awful Throne, that Paradice….
> (44-47)

Perhaps if Lysander had continued to regard Cloris in pieces, he could have accomplished what they both seemed to want, and, further, perhaps he could have "set his seal" on the "altar" that is not yet his with the same vigorous efficiency that motivated Donne and Vesalius. But he doesn't. When Cloris discreetly succumbs—"And now no signs of Life she shows,/ but what in short-breath'd Sighs returns & goes"—Lysander makes the mistake of looking at her as a unified and dazzling whole, an equal self, giving that self to him:

> He saw how at her Length she lay,
> He saw her rising bosom bare;
> Her loose thin *Robes*, through which appear

> A Shape design'd for Love and Play....
> Off'ring her Virgin-Innocence.
>
> (61-4, 67)

Confronted by equally desiring desire, Lysander is just not up to it. Behn's final joke at the expense of both poets and anatomists is that a man like Lysander can mount the "awful throne" only if he sees it as a body part. If he lets himself gaze on it as the central secret in a living, equally unified individual, the effect is emasculating and insur(mount)able. Lysander is desperate. He throws himself on Cloris, completing his dissection by tearing off the last of her "willing garments," but it is too late; he tries to work up his enthusiasm all by himself—but to no avail. He is a victim of his own anatomical gaze.

Sawday, in his discussion of the dissection of female bodies, alludes to the fact that those first early modern anatomists had to face "a fear of interiority; more often than not, specifically male fear of the female interior," and he argues that "once the body has been partitioned and its interior dimensions laid open to scrutiny, the very categories 'male' and 'female' become fluid, even interchangeable."[39]

First, of course, Behn is suggesting that the parts of a woman really are much more than some of the whole; more interestingly, with the aid of traditional humoral theory, she problematizes our notions of sexuality in a way that wasn't really possible again until the advent of a post modern queer theory developed by thinkers like Michele Foucault, Eve Kosofsky Sedgwick, Jeffrey Weeks, Jonathan Goldberg, Kate Bornstein and Judith Butler. In her poetry Behn constantly manipulates sexual categories to show how easily elisions of male and female sexuality can be accomplished. In this poem, Behn, in addition to anatomically exposing male anxiety about equal female desire, uses Galenic humoral theory to "feminize" Lysander, to "masculinize" Cloris, and once again to destabilize the notion of sexual category.

---

[39] Sawday, 3.

Roberta C. Martin (East Carolina University) is Associate Professor of English in the Department of English at East Carolina University in Greenville, North Carolina. She teaches and researches topics in Milton studies, seventeenth century women, and the history of sexuality. She has published articles on Milton and Aphra Behn.

BIBLIOGRAPHY

Aberdeen Bestiary. English 1200. Aberdeen University Library MS 24, folio 70r.

Adams, Frank Dawson. *The Birth and Development of the Geological Sciences*. New York: Dover, 1954.

Agricola, Georgius. *De Re Metallica*. Herbert Clark Hoover and Lou Henry Hoover (trans.). New York: Dover, 1950.

Albanese, Denise. *New Science, New World*. Durham: Duke University Press, 1996.

Aldrovandi, Ulisse Ambrosini Bartolommeo. *Vlyssis Aldrovandi Patricii Bononiensis De Quadrupedibus Digitatis Viviparis Libri Tres, Et De Quadrupedibus Digitatis Oviparis Libri Duo*. Bologna: Apud Nicolaum Tebaldinum, 1637.

Aldrovandi, Ulisse and Ambrosini Bartolommeo. *Vlyssis Aldrovandi Patricii Bononiensis Serpentum, Et Draconum Historiae Libri Duo*. Bologna: Apud Clementem Ferronium, 1640.

Allison, Henry (ed. and trans.). *The Kant-Eberhard Controversy*. Baltimore: Johns Hopkins University Press, 1973.

Arendt, Hannah. *The Human Condition*. Garden City, NJ: Doubleday-Anchor, 1959.

Asch, Ronald G. *The Thirty Years War: The Holy Roman Empire and Europe, 1618-48*. New York: St. Martin's Press, 1997.

Ashworth Jr., William B. "Marcus Gheeraerts and the Aesopic Connection in Seventeenth-Century Scientific Illustration." *Art Journal* 44:2 (1984), 135.

Asuá, Miguel and Roger French. *A New World of Animals: Early Modern Europeans on the Creatures of Iberian America*. Aldershot, England: Ashgate, 2005.

Bacon, Francis. *Advancement of Learning*. Michael Kiernan (ed.). New York: Oxford University Press, 2000.

_____. *New Atlantis*. Kila: Kessinger, 1997.

_____. *New Organon*. Peter Urbach and John Gibson (trans.). C hicago: Open Court, 1994.

_____. *Sylva Sylvarum*. (trans.). Basil Montagu. London: William Pickering, 1826.

_____. *Wisdom of the Ancients*. Kila: Kessinger, 1997.

Bayle, Pierre. *Dictionnaire historique et critique*. 5th ed., 4 vols. Amsterdam: P. Brunel et. al., 1740.

Behn, Aphra. *Poetry*. Vol. 1, *The Works of Aphra Behn*. Janet Todd, (ed.). Columbus: Ohio State University Press, 1992-96.

Bell, Daniel. "Herping the Night Sky: The Mythology of Snake and Lizard Constellations." *Sonoran Herpetologist* 18:6 (2005), 62-65.

Belon, Pierre. *Les Observations De Plusieurs Singularités Et Choses Mémorables, Trouvées En Grèce, Turquie, Judée, Egypte, Arabie & Autres Pays Étranges.* Paris: Guillaume Cauellat, 1554.

Bennett, Jonathan. *A Study of Spinoza's 'Ethics.'* Indianapolis: Hackett, 1984.

Bloch, Ernst. *Natural Law and Human Dignity.* Cambridge, MA: MIT Press, 1988.

Boswell, Jackson Campbell. *Milton's Library: A catalogue of the remains of John Milton's library and an annotated reconstruction of Milton's library and ancillary readings.* New York: Garland Publishers, 1975.

Breitenberg, Mark. *Anxious Masculinity in Early Modern England.* Cambridge: Cambridge University Press, 1996.

Brown, Harcourt. "History and the Learned Journal." *Journal of the History of Ideas* 33:3 (Jul.-Sept. 1972): 365-78.

Bry, Theodor de. *Brevis Narratio Eorvm Qvæ in Florida Americæ Provicia Gallis Acciderunt, Secunda in Illam Nauigatione, Duce Renato De Laudoniere ... Anno Mdlxiiii. Qvae Est Secvnda Pars Americae... Auctore Iacobo Le Moyne, Cui Cognomen De Morgues.* Frankfurt: I.Wecheli, 1591.

Buffon, Georges Louis Leclerc comte de. *Histoire Naturelle, Générale Et Particulière Avec La Description Du Cabinet Du Roi.* Paris: De l'Imprimerie royale, 1749-1804.

Bull, Hedley, Benedict Kingsbury and Adam Roberts, (eds.). *Hugo Grotius and International Relations.* Oxford: Clarendon Press, 1990.

Burnet, Thomas. *The Sacred Theory of the Earth, Containing an Account of the Original of the Earth and of all the General Changes which it hath already undergone, or is to undergo 'till the Consumation of all Things.* London: R. Norton, 1684.

Cadden, Joan. *Meanings of Sex Difference in the Middle Ages: Medicine, Science, and Culture.* New York: Cambridge University Press, 1993.

Catesby, Mark. *The Natural History of Carolina, Florida and the Bahama Islands: Containing the Figures of Birds, Beasts, Fishes, Serpents, Insects, and Plants; Particularly the Forest-Trees, Shrubs, and Other Plants, Not Hitherto Described, or Very Incorrectly Figured by Authors. Together with Their Descriptions in English and French. To Which, Are Added Observations on the Air, Soil, and Waters; with Remarks Upon Agriculture, Grain, Pulse, Roots, Etc. To the Whole, Is Prefixed a New and Correct Map of the Countries Treated of, 2 Volumes.* London: Published by the author, 1731-1743.

Chandler, James, Arnold Davidson, and Adrian Johns (eds.). "Arts of Transmission." *Critical Inquiry* 31:1 (August 2004).

Chartier, Roger. *The Order of Books: Readers, Authors, and Libraries in Europe between the Fourteenth and Eighteenth Centuries*. Lydia G. Cochrane (trans.). Stanford: Stanford University Press, 1994.

Clements, Arthur L. (ed.). *John Donne's Poetry: Authoritative texts, Criticism.* Norton Critical Editions. New York: Norton, 1992.

Colebrook, Claire. *New Literary Histories: New Historicism and Contemporary Criticism*. Manchester: Manchester University Press, 1997.

Collins, Minta. *Medieval Herbals: The Illustrative Traditions*. Toronto: University of Toronto Press, 2000.

Conrad, Lawrence I. "The Arab-Islamic Medical Tradition." In Conrad, *The Western Medical Tradition: 800 BC to AD 1800*, 93-138.

Conrad, Lawrence I. et al, (ed.). *The Western Medical Tradition: 800 BC to AD 1800*. Cambridge: Cambridge University Press, 1995.

Corominas, Joan and José A. Pascual. *Diccionario Crítico Etimológico Castellano e Hispánico*. Madrid: Gredos, 1984.

Covarrubias, Sebastián de. *Tesoro de la Lengua Castellana o Española*. Martín de Riquer (ed). Barcelona: S.A. Horta, 1943.

Crawford, Patricia. "Sexual Knowledge in England, 1500-1700." In *Sexual Knowledge, Sexual Science: the History of Attitudes toward Sexuality.* Roy Porter and Mikulas Teich, (eds). Cambridge: Cambridge University Press, 1994.

Cruz-Sotomayor, Beatriz. "Una voz en busca de autorización: Oliva Sabuco y su filosofía médica de la felicidad." M.A. thesis, University of Puerto Rico, 2001.

Darwin, Charles. *Journal of Researches into the Geology and Natural History of the Various Countries Visited by H. M. S. Beagle, under the Command of Captain Fitz Roy, R. N., from 1832-1836*. London: Henry Colburn, 1839.

Descartes, René. *Oeuvres de Descarte.* 11 vols. Charles Adam and Paul Tannery (eds.). Paris: Vrin, 1996.

Donelan, Michael. "Grotius and the Image of War," *Millennium* 12 (1983): 233-43.

Donne, John. *Poetical Works*. (ed). Sir Herbert Grierson. London: Oxford University Press, 1933.

_____. *The Complete English Poems.* A. J. Smith (ed.). New York: Penguin, 1996.

Dunlop, Ian. *The Palaces and Progresses of Elizabeth I.* London: Cape, 1962.

Durant, David. *Bess of Hardwick: Portrait of an Elizabethan Dynasty*. London: Weidenfeld & Nicolson, 1977; reprint 1998.

Edwards, Charles S. *Hugo Grotius, The Miracle of Holland.* Chicago: Nelson-Hall, 1981.

Egerton, Frank N. "A History of the Ecological Sciences, Part 11: Emergence of Vertebrate Zoology during the 1500s." *Ecological Society of America Bulletin* 84 (2003), 206–12.

Eisenstein, Elizabeth. *The Printing Press as an Agent of Change: Communications and Cultural Transformations in Early Modern Europe*. Cambridge: Cambridge University Press, 1979.

Epicurus. *The Epicurus Reader*. Brad Inwood and L. P. Gerson (eds. and trans.). Indianapolis: Hackett Publishing, 1994.

Evans, J. Martin. *Paradise Lost and the Genesis Tradition*. Oxford: Clarendon Press, 1968.

_____. *The Miltonic Moment*. Lexington, KY: The University Press of Kentucky, 1998.

Ferrari, Giovanna. "Public Anatomy Lessons and the Carnival: The Anatomy Theatre of Bologna." *Past and Present* 117, (November 1987): 50-106.

Ferrone, Vincenzo. *The Intellectual Roots of the Italian Enlightenment*. Sue Brotherton (trans.). Atlantic Highlands: Humanities Press, 1995.

Findlen, Paula. *Possessing Nature: Museums, Collecting, and Scientific Culture in Early Modern Italy*. Berkeley: University of California Press, 1994.

Freeman, James A. *Milton and the Martial Muse, Paradise Lost and European Traditions of War*. Princeton: Princeton University Press, 1980.

French, Roger. *Medicine before Science: The Business of Medicine from the Middle Ages to the Enlightenment.* Cambridge: Cambridge University Press, 2003.

French, Sara. "Women, Space and Power: The Building and Use of Hardwick Hall in Elizabethan England." Ph.D. diss., Binghamton University, 2000.

Frick, George and Raymond Stearns. *Mark Catesby: The Colonial Audubon.* Urbana: University of Illinois Press, 1961.

Friedman, Alice. "Architecture, Authority and the Female Gaze: Planning and Representation in the Early Modern Country House." *Assemblage* 18 (August 1992):41-61.

_____. *House and Household in Elizabethan England: Wollaton Hall and the Willoughby Family*. Chicago: University of Chicago Press, 1989.

Gandelman, Claude. "The Poem as Map: John Donne and the 'Anthropo-morphic Landscape' Tradition." *Applied Semiotics / Sémiotique appliquée* 3:8 (1999): 207-215. Published originally in *Arcadia: Zeitschrift für VergleichendeLiteraturwissenschaft*, 19 (1984): 244-51.

Garber, Daniel, *Descartes Embodied: Reading Cartesian Philosophy Through Cartesian Science*. Cambridge: Cambridge University Press, 2001.

García Rubio, Samuel. "La obra latina del bachiller Sabuco." *Al-Basit. Revista de Estudios Albacetenses. Monográfico dedicado a Miguel Sabuco* 22 (Dec. 1987): 217-232.

Generali, Dario. "Fra Filologia e storiografia: Progetti di edizioni settecentesche." *StudiFilosofici* 18 (1995): 157-172.

George, Wilma B. and Brunsden Yapp. *The Naming of the Beasts: Natural History in the Medieval Bestiary*. London: Duckworth, 1991.

Gesner, Konrad. *Historia Animalium Liber II: De Quadrupedibus Ouiparis*. Frankfurt: Roberti Cambieri, 1586.

_____. *Historia Animalium. Liber V: Qui Est De Serpentium Natura.* Zurich: Froschouiana 1587.

Ginzburg, Carlo. *The Cheese and the Worms: The Cosmos of a Sixteenth-Century Miller.* John and Anne Tedeschi (trans.). New York: Penguin, 1980.

_____. "Micro History: Two or Three Things That I Know About It." John Tedeschi and Anne C. Tedeschi (trans.). *Critical Inquiry* 20:1 (Autumn 1993), 10-35.

Girouard, Mark. *Life in the English Country House: A Social and Architectural History.* New Haven: Yale University Press, 1978.

_____. *Robert Smythson and the Elizabethan Country House.* New Haven: Yale University Press, 1983.

Gould, Stephen Jay. *Time's Arrow, Time's Cycle: Myth and Metaphor in the Discovery of Geological Time.* Cambridge, MA: Harvard University Press, 1987.

Grafton, Anthony. *Cardano's Cosmos: The Worlds and Works of a Renaissance Astrologer.* Cambridge, MA: Harvard University Press, 1999.

_____. *Commerce With the Classics: Ancient Books and Renaissance Readers.* Ann Arbor: University of Michigan Press, 1997.

Granjel, Luis S. "El jugo nérveo." In *La doctrina antropológico-médica de Miguel Sabuco.* Salamanca: Universidad de Salamanca, 1956. 43-50.

Green, Monica, (ed. and trans.). *The Trotula. A Medieval Compendium of Women's Medicine.* Philadelphia: University of Pennsylvania Press, 2001.

Greenblatt, Stephen. *Renaissance Self-Fashioning: From Moore to Shakespeare.* Chicago: University of Chicago Press, 1980.

Greenblatt, Stephen and Catherine Gallagher. *Practicing New Historicism.* Chicago: University of Chicago Press, 2000.

Grotius, Hugo. *De Jure Belli ac Paci Libri Tres.* (trans.). Francis W. Kelsey. London: Wildy & Sons, 1964.

_____. *De Jure Praedae Commentarius.* Gwladys L. Williams (trans.). London: Wildy & Sons, 1950.

Gudger, E.W. "The Five Great Naturalists of the Sixteenth Century: Belon, Rondelet, Salviani, Gesner and Aldrovandi: A Chapter in the History of Ichthyology." *Isis* 22:1 (1934), 21-40.

Haines, Sally. *Slithy Toves: Illustrated Classic Herpetological Books at the University of Kansas in Pictures and Conversations.* Ithaca, NY: Society for the Study of Amphibians and Reptiles 2000.

Hale, J.R. *Renaissance War Studies.* London: The Hambledon Press, 1983.

_____. *War and Society in Renaissance Europe, 1450 -1620.* New York: St. Martin's Press, 1985.

Hall, Bert S. *Weapons and Warfare in Renaissance Europe: Gunpowder, Technology, and Tactics.* Baltimore & London: The Johns Hopkins University Press, 1997.

Hanford, James Holly. *A Milton Handbook.* 4th ed. New York: Appleton-

Century-Crofts, 1961.
Harvey, William. *The Anatomical Exercises: De Motu Cordis and De Cirulatione Sanguinis in English Translation.* Geoffrey Keynes (ed.). New York: Dover, 1995. Reprint of *The Anatomical Exercises of Dr. William Harvey, De Motu Cordis 1628; De Circulatione Sanguinis, 1649.* London: Nonesuch Press 1953.
Healy, Thomas and Jonathan Sawday (eds.). *Literature and the English Civil War.* Cambridge: Cambridge University Press, 1990.
Hitchcock, Tim. *English Sexualities, 1700-1800.* New York: St. Martin's Press, 1997.
Horne, Charles F. *Medieval Arabia.* Vol. 6, *The Sacred Books and Early Literature of the East.* New York: Parke, Austin, & Lipscomb, 1917.
Howard, Michael. *The Causes of War.* Cambridge, MA: Harvard University Press, 1984.
Huarte de San Juan, Juan. *Examen de ingenios para las ciencias.* 1575. In *Biblioteca de Autores Españoles. Desde la formación del lenguaje hasta nuestros días. Obras escogidas de filósofos.* Adolfo de Castro (ed.) Madrid: RAE, 1953. 397-520.
_____. *Examen de ingenios. The Examination of Men's Wits (1594) by Juan Huarte.* Translated into Italian by Camillo Camilli and then into English by Richard Carew. A facsimile reproduction. Gainesville, Florida: Scholar's Facsimiles & Reprints, 1959. Originally published in London: Adam Flip, 1594.
Hulton, Paul. *The Work of Jacques Le Moyne De Morgues: A Huguenot Artist in France, Florida, and England.* London: British Museum Publications, 1977.
Imperato, Ferrante. *Dell'historia Naturale.* Napoli: C. Vitale, 1599.
Islamic Culture and the Medical Arts: Late Medieval and Early Modern Medicine. http://www.nlm.nih.gov/exhibition/islamic_medical/islamic_14.html. (February, 2007).
Israel, Jonathan. *Radical Enlightenment: Philosophy and the Making of Modernity, 1650-1750.* Oxford: Oxford University Press, 2001.
Jacob, James R. *The Scientific Revolution: Aspirations and Achievements, 1500-1700.* Atlantic Highlands, NJ: Humanities Press, 1998.
Jacobaeus, Oliger. *De Ranis Et Lacertis Observationes.* Copenhagen: Impensis Johannis M. Lieben, 1686.
Jardine, Nicholas, James A. Secord and E. C. Spary. *Cultures of Natural History.* Cambridge: Cambridge University Press, 1996.
Johns, Adrian. *The Nature of the Book: Print and Knowledge in the Making.* Chicago: University of Chicago Press, 1998.
Jonstonus, Joannes. *Historiae Naturalis.* Amstelodami: Apud Ioannem Iacobi fil. Schipper, 1657.

Jordan, Constance. *Renaissance Feminism: Literary Texts and Political Models.* Ithaca, NY: Cornell University Press, 1990.

Kamps, Ivo (ed.). *Materialist Shakespeare: A History.* London: Verso, 1995.

Kant, Immanuel. *Werkausgabe.* 12 vols. Wilhelm Weischedel (ed.). Frankfurt am Main: Suhrkamp, 1996.

Keller, Evelyn Fox. *Reflections on Gender and Science.* New Haven: Yale University Press, 1985.

Kelly, Jack. *Gunpowder: alchemy, bombards, and pyrotechnics--The history of the explosive that changed the world.* New York: Basic Books, 2004.

Kray Jill (ed.). *Classical Traditions in Renaissance Philosophy.* Aldershot: Ashgate, 2002.

Laqueur, Thomas. *Making Sex: Body and Gender from the Greeks to Freud.* Cambridge, MA: Harvard University Press, 1990.

Lauterpacht, Hersh. *International Law and Human Rights.* New York: F. A. Praeger, 1950.

Le Doeff, Michèle. *The Philosophical Imaginary.* Stanford: Stanford University Press, 1989.

Lee, Stephen J. *The Thirty Years War.* London: Routledge, 1991.

Leibniz, Gottfried Wilhelm. *Die philosophischen Schriften.* Vol. 3. Berlin: Herausgegeben, 1875-1890.

_____. *Die Philosophische Schriften.* 7 vols. C. J. Gerhardt (ed.). Hildesheim: Olms, 1978.

_____. *Protogaea.* Goettingen: C. L. Scheidio, 1749.

_____. *Leibniz's 'New System' and Associated Contemporary Texts.* R. S. Woolhouse and Richard Francks (eds. and trans.). Oxford: Oxford University Press, 1997.

_____. "Système nouveau de la nature et de la communication des substances, aussi bien que de l'union qu'il y a entre l'âme et le corps." *Le Journal des savants,* Paris ed. 25-26 (June 27 and July 4, 1695), 294-306.

Lennon, Thomas. *The Battle of the Gods and Giants: The Legacies of Descartes and Gassendi,1655-1715.* Princeton: Princeton University Press, 1993.

_____. *Reading Bayle.* Toronto: University of Toronto Press, 1999.

Levin, Carole. *The Heart and Stomach of a King: Elizabeth I and the Politics of Sex and Power.* Philadelphia: University of Pennsylvania Press, 1994.

Lindemann, Mary. *Medicine and Society in Early Modern Europe.* Cambridge: Cambridge University Press, 1999.

Loades, David. *The Tudor Court.* London: B. T. Batsford, 1992.

Lorber, Judith. *Paradoxes of Gender.* New Haven: Yale, 1994.

Lovejoy, Arthur O. *The Great Chain of Being: A Study of the History of an Idea.* New York: Harper, 1960.

Maillet, Benoît de. *Telliamed.* Vol. 2. Amsterdam: Chez L'honore & Fils, 1748.

Martín-Araguz, A., C. Bustamante-Martínez, and V. Fernández-Armayor. "El suco nérveo sabuceano y los orígenes de la neuroquímica en el Renacimiento español." *Revista de Neurología* 36:12 (2003): 1190-1198.

Mazzolini, Renato G. *Non-Verbal Communication in Science Prior to 1900.* Florence: L.S. Olschki, 1993.

McKeon, Michael. "Historicizing Patriarchy: The Emergence of Gender Difference in England, 1660-1760." *Eighteenth-Century Studies* 28:3 (1995): 295-322.

Mendelson, Sara and Patricia Crawford. *Women in Early Modern England, 1550-1720.* Oxford: Clarendon Press, 1998.

Mercer, Christia and Eileen O'Neill (eds.) *Early Modern Philosophy: Mind, Matter, and Metaphysics.* Oxford: Oxford University Press, 2005.

Merchant, Carolyn. *The Death of Nature: Women, Ecology, and the Scientific Revolution.* San Francisco: Harper and Row, 1980.

Merian, Maria Sibylla. *Metamorphosis Insectorum Surinamensium.* Amsterdam: Published by the author, 1705.

_____. *Metamorphosis Insectorum Surinamensium.* Amsterdam: Joannem Oosterwyk, 1719.

_____. *Schmetterlinge, Kăfer Und Andere Insekten: Leningrader Studienbuch.* Wolf Beer (ed.). Luzern: Reich, 1976.

Merian, Maria Sibylla, William T. Stearn, Vitor Osmar Becker and Elisabeth Rücker. *Metamorphosis Insectorum Surinamensium.* London: Pion, 1980.

Meyers, Amy and Margaret Pritchard. *Empire's Nature: Mark Catesby's New World Vision.* Chapel Hill: University of North Carolina Press, 1998.

Milner, Andrew. *John Milton and the English Revolution, A Study in the Sociology of Literature.* London: Macmillan Press, 1981.

Milton, John. *Complete Prose Works of John Milton.* 8 vols. Don M. Wolfe et al (eds.). New Haven, CT: Yale University Press, 1953-83.

_____. *The Poetical Works of John Milton.* Douglas Bush (ed.). London: Oxford University Press, 1966.

_____. *The Poems of John Milton.* John Carey and Alastair Fowler (eds.). Longman: London, 1968.

_____. *The Works of John Milton.* Frank Allen Patterson (ed.). New York: Columbia University Press. 1931.

Moliner, María. *Diccionario del Uso del Español.* Madrid: Gredos, 1998.

Montalenti, Giuseppe. "Antonio Vallisneri." *Dictionary of Scientific Biography* 13 (1976): 562-5.

Moureau, François (ed.). *Les Presses grises: La Contrefaçon du livre.* Paris: Aux amateurs de livres, 1988.

Mumford, Lewis. *Technics and Civilization.* New York: Harcourt, Brace & World, 1962.

Muratori, L. A. *Carteggi con Ubaldini...Vannoni.* In *Edizione nazionale del Carteggio di L.A. Muratori.* Michela L. Nichetti Spanio (ed.). Vol. 44. Florence: Olschki, 1978.

Nutton, Vivian. "The Fatal Embrace: Galen and the History of Ancient Medicine." *Science in Context* 18:1 (2005): 11-121.

_____. "Medicine in the Greek World, 800-50 BC." In Conrad, *The Western Medical Tradition*, 11-38.

_____. "Medicine in Late Antiquity and the Early Middle Ages." In Conrad, *The Western Medical Tradition*, 71-87.

_____. "Roman Medicine, 250 BC to AD 200." In Conrad, *The Western Medical Tradition*, 39-70.

Öhman, Arne and Susan Mineka. "The Malicious Serpent: Snakes as a Prototypical Stimulus for an Evolved Module of Fear." *Current Directions in Psychological Science* 12:1 (2003), 5-9.

Oldroyd, David R. *Thinking About the Earth: A History of Ideas in Geology*. Cambridge, MA: Harvard University Press, 1996.

Orlin, Lena Cowen. "'The Causes and Reasons of All Artificial Things' in the Elizabethan Domestic Environment." *Medieval and Renaissance Drama in England* 7 (1995): 19-75.

_____. *Private Matters and Public Culture in Post-Reformation England.* Ithaca, NY: Cornell University Press, 1994.

Osler, Margaret J. "The *History* of Philosophy and the *History* of Philosophy: A Plea for Textual History in Context." *Journal of the History of Philosophy* 40:4 (Oct. 2002), 529-33

Otero-Torres, Damaris. "'Una humilde sierva osa halbar' o la ley del padre; dislocaciones entre texto femenino y autoría masculina en 'La carta introductoria al Rey Nuestro Señor' de Oliva Sabuco de Nantes." *Taller de Letras* 26: (1998): 9-27.

Parker, Geoffrey (ed.). *The Thirty Years War*. London: Routledge & Kegan Paul, 1984.

Paxson, James J. "The Green[blatt]ing of America: Reflections on the Institutional Genealogy of Greenblatt's New Historicism." *The Minnesota Review* n.s. 41-42 (1993), 221-35.

Piso, Willem and Georg Marggraf. *Historia Naturalis Brasiliae*. Amsterdam: Franciscum Hackium, 1648.

Pliant, Jack. "'Pleading the Belly'[:] Angela Barnett Avoid[s] 18th Century Death Sentence." (2005). geneologymagizine.com. http://www.genealogymagazine.com/plbeanbaav18.html. (February, 2007).

Pliny the Elder. *The Natural History of Pliny*. John Bostock and H.T. Riley (trans.). Vol 2. London: Henry G. Bohn, 1855.

Popkin, Richard. "Shadow History." *Journal of the History of Philosophy* 31:1 (Jan. 1993), 119-22.

_____. *The Third Force in Seventeenth-Century Thought*. Leiden: Brill, 1992.

Porqueras Mayo, Alberto. *El prólogo en el Renacimiento español*. Madrid: Consejo Superior de Investigaciones Científicas, 1965.

Porter, Roy. *The Making of Geology: Earth Science in Britain 1660-1815*. Cambridge: Cambridge University Press, 1977.

Ramoutsaki, I. A., S. Haniotakis and A. M. Tsatsakis. "The Snake as the Symbol of Medicine, Toxicology and Toxinology." *Veterinary and Human Toxicology* 42:5 (2000), 306-8.

Rappaport, Rhoda. "Italy and Europe: the case of Antonio Vallisneri (1661-1730)." *History of Science* 29 (1991): 73-98.

_____. *When Geologists Were Historians, 1665-1750*. Ithaca, NY: Cornell University Press, 1997.

Rodríguez de la Torre, Fernando. "El autor y la autoría en la obra de Sabuco." *Al-Basit. Revista de Estudios Albacetenses. Monográfico dedicado a Miguel Sabuco* 22: (Dec. 1987): 233-265.

Rorty, Richard, J. B. Schneewind, and Quentin Skinner (eds.). *Philosophy in History: Essays on the Historiography of Philosophy*. Cambridge: Cambridge University Press, 1984.

Rösel von Rosenhof, August Johann. *Der Monatlich-Herausgegebenen Insecten-Belustigung*. Nuremberg: Johann Joseph Fleischmann, 1746.

_____. *Historia Naturalis Ranarum Nostratium: In Qua Omnes Earum Proprietates, Praesertim Quae Ad Generationem Ipsarum Pertinent, Fusius Enarrantur*. Nuremberg: Johann Joseph Fleischmann, 1758.

Rossi, Paolo. *The Dark Abyss of Time*. Lydia G. Cochrane (trans.). Chicago: University of Chicago Press, 1984.

_____. *Francis Bacon: From Magic to Science*. (trans.). Sacha Rabinovitch. Chicago: University of Chicago Press, 1968.

Rublack, Ulinka. "Fluxes: the Early Modern Body and the Emotions." (trans.). Pamela Selwyn in *History Workshop Journal* 53 (2002): 1-16.

Rudwick, Martin J. S. *The Meaning of Fossils: Episodes in the History of Paleontology*. New York: Neal Watson, 1976.

Russell, Bertrand. *A Critical Exposition of the Philosophy of Leibniz*. London: Routledge, 1992.

Sabuco de Nantes, Oliva. *Obras de doña Oliva Sabuco de Nantes y Barrera (Escritora del siglo XVI)*. 1587. Octavio Cuartero (ed.). Madrid: Establecimiento Tipográfico de Ricardo Fe, 1888.

_____. *New Philosophy of Human Nature Neither Known to Nor Attained by the Great Ancient Philosophers, Which Will Improve Human Life and Health*. Mary Ellen Waithe, María Colomer Vintró, and Ángel Zorita (trans.). Illinois: University of Illinois Press, 2007.

Sawday, Jonathan. *The Body Emblazoned: Dissection and the Human Body in Renaissance Culture*. New York: Routledge, 1995.

Shakespeare, William. *Macbeth*. Kenneth Muir (ed.). *The Arden Edition of the Works of William Shakespeare*. Richard Proudfoot (ed.). 35 vols. New York: Methuen, 1984. Reprint, Methuen, 1986.

Shapin, Steven. "Of Gods and Kings: Natural Philosophy and Politics in the Leibniz-Clarke Disputes." *Isis* 72 (1981): 187-215.

Siraisi, Nancy G. "Vesalius and the Reading of Galen's Teleology." *Renaissance Quarterly* 50:1 (Spring 1997): 1-37.

Solomon, Julie. *Objectivity in the Making: Francis Bacon and the Politics of Inquiry*. Baltimore: Johns Hopkins University Press, 1997.

Soubani, Ayman O. and Faroque A. Khan. "The Discovery of the Pulmonary Circulation Revisited." Annals of Saudi Medicine. (March, 1995), http://www.kfshrc.edu.sa/annals/152/mh9422ar.html. (January, 2007).

Starkey, David. "Intimacy and Innovation: The Rise of the Privy Chamber, 1485-1547." *The English Court: From the Wars of the Roses to the Civil War*. Harlow, England: Longman, 1987: 71-118.

Stearn, William T. "Maria Sibylla Merian (1647-1717) as a Botanical Artist." *Taxon* 31:3 (1982), 529-34.

Stone, Gilbert. *The British Coal Industry*. London: J. M. Dent & Sons. 1919.

Summerson, John. *Architecture in Britain 1530-1830*. 9th ed. New Haven: Yale University Press, 1993.

Thomas, Keith. *Man and the Natural World: Changing Attitudes in England, 1500-1800*. New York: Oxford University Press, 1996.

Todd, Janet. *The Secret Life of Aphra Behn*. New Brunswick: Rutgers University Press, 1997.

Topsell, Edward. *The History of Four-Footed Beasts and Serpents: Describing at Large Their True and Lively Figure, Their Several Names, Conditions, Kinds, Virtues (Both Natural and Medicinal), Countries of Their Breed, Their Love and Hatred to Mankind, and the Wonderful Work of God in Their Creation, Preservation, and Destruction*. London: Printed by E. Cotes, for G. Sawbridge ..., T. Williams ..., and T. Johnson ... 1658.

Trevor-Roper, Hugh. *Catholics, Anglicans, and Puritans, Seventeenth-Century Essays*. Chicago: University of Chicago Press, 1988.

_____. *From Counter-Reformation to Glorious Revolution*. Chicago: University of Chicago Press, 1992.

Tucci, Francesco Saverio. "Il parlare della S. Scrittura e l'operare della natura: gli interrogativi della geologica storica nella riflessione di Antonio Vallisneri." *Contributi* 7 (1983), 27.

Vallisneri, Antonio. *De Corpi marini che su Monti si trovano...* Venice: Lovisa, 1721.

_____. *Opere diverse... I. Istoria del Camaleonte Affricano, e di vari Animali d'Italia... II. Lezione Accademia intorno l'Origine delle Fontane, Colle Annotazioni per chiarezza maggiore della medesima... III. Raccolta di vari Trattati*. Venice : G. Hertz, 1715.

Van De Graaff, Kent M. *Human Anatomy*. Columbus, Ohio: McGraw-Hill, 2001.

Veeser, H. Aram (ed.). *The New Historicism*. London: Routledge, 1989.

_____, H. Aram (ed.). *The New Historicism Reader*. London: Routledge, 1994.

Watson, Richard A. "Shadow History in Philosophy." *Journal of the History of Philosophy* 31:1 (Jan. 1993), 95-109.

Wear, Andrew. "Medicine in Early Modern Europe, 1500-700." In Conrad, *The Western Medical Tradition*, 215-361.

Webster's New Twentieth Century Dictionary of the English Language. Unabridged. New York: William Collins, 1980.

Wedgwood, C. V. *The Thirty Years War.* London, New York: Metheun, 1981.

Whitney, Charles. *Francis Bacon and Modernity.* New Haven: Yale University Press, 1986.

Wigley, Mark. "Untitled: The Housing of Gender" in *Sexuality & Space.* Beatriz Colomina (ed.). Princeton: Princeton University Press, 1992: 327-389.

Williams, Rosalind. *Notes on the Underground: An Essay on Technology, Society, and the Imagination.* Cambridge, MA: MIT Press, 1990.

Wilton-Ely, John. "The Professional Architect in England," *The Architect: Chapters in the History of the Profession.* Spiro Kostof (ed.). New York: Oxford University Press, 1977; Oxford University Press paperback, 1986: 180-208.

Wolfe, Jessica. *Humanism, Machinery, and Renaissance Literature.* New York: Cambridge University Press, 1994.

Woodward, John. *An Essay toward a Natural History of the Earth: and Terrestrial Bodies, Especially Minerals: As also of the Sea, Rivers, and Springs, With an Account of the Universal Deluge: And of the Effects that it had upon the Earth.* London: Richard Wilkin, 1695.

Woolgar, C. M. *The Great Household in Late Medieval England.* New Haven: Yale University Press, 1999.

Worster, Donald. *Nature's Economy: The Roots of Ecology.* San Francisco: Sierra Club, 1977.

Wright, Pam. "A Change in Direction: The Ramifications of a Female Household, 1558-1603," *The English Court: From the Wars of the Roses to the Civil War.* Harlow, England: Longman, 1987: 147-172.